Praise for Third Edition of *Hot Talk, Cold Science*

"In *Hot Talk, Cold Science*, Fred Singer looks at the issue of climate change the way a physicist should. He asks probing questions and offers reasoned possibilities. He notes the obvious weaknesses that others too often ignore and freely acknowledges his own limitations. The biggest weakness is the obvious refusal of those promoting alarm to allow the questioning that is the heart of science. Nothing better illustrates the fact that we are dealing with a political *cum* religious cult rather than science where the quest for power overwhelms scientific inquiry. Alas, even scientists are often attracted by power and public recognition. Fortunately, some like Dr. Singer still prefer the joys and value of scientific inquiry."

 —**Richard S. Lindzen**, Alfred P. Sloan Professor Emeritus of Meteorology; Department of Earth, Atmospheric and Planetary Sciences; M.I.T.

"As contentious as the climate issue was and is, I was always impressed by Fred Singer's gentle demeanor within that storm. I suppose he remained calm because he sought to ground his views in the actual evidence of climate observations. In his day, reproducible evidence was the foundation on which one was taught to test one's claims and he simply went about the business of checking out the latest theorized conjectures about the climate. Here in *Hot Talk, Cold Science*, he updates some of his findings regarding those conjectures, as well as giving a little tour of the political landscape that melded itself to the climate-alarm agenda. His conclusions should give us all a modest sense of gentle calmness—that same calmness he carried to the end of his days."

 —**John R. Christy**, Distinguished Professor of Atmospheric Science; Interim Dean, College of Science; Director, Earth System Science Center, Johnson Research Center, University of Alabama in Huntsville; Alabama State Climatologist

"When debating environmental policy, we frequently hear from scientists, climate activists, and public officials who claim 'the science is settled' with regard to global warming. This book is a great reminder that the data are mixed at best. We should follow the science. *Hot Talk, Cold Science* provides the reader with important facts and evidence consistently and conveniently overlooked by climate alarmists, making clear the case on global warming is far from closed."

 —**Ted Cruz**, U.S. Senator; Chairman, Senate Subcommittee on Science and Space; Chairman, Senate Subcommittee on the Constitution, Civil Rights and Human Rights

"*Hot Talk, Cold Science* is an excellent book on the politics and science of climate change. I particularly enjoyed the first part, 'Hot Talk,' that shows the duplicity of the climate alarmist community in exaggerating claims of imminent disaster (that has been 10 years away for decades) as well as suppressing contrary opinions from highly respected scientists ('cancel culture'). I did not realize the true extent of these propaganda efforts until I read this book, which documents very well many of these shenanigans. The general public should know how far the alarmist 'climate scientists' (including Al Gore) strayed from good science in trying to bamboozle them. I have also found the second part 'Cold Science' enlightening, particularly the sections discussing how poorly the IPCC climate models have 'predicted' climate measurements. An analogous situation is the recent case of almost total failure of the epidemiological models to predict the COVID-19 pandemic's progression and resulting deaths. In the case of the epidemiological models the general public can see almost in real time how poorly predictive they are. In both cases, one is trying to solve a non-linear, chaotic system to predict its future. It is well known in mathematics and physics that one does very poorly in predicting the detailed future time dependence of such a system. Yet the climate modelers persist in their folly, basing their models on a linear cause and effect of increasing anthropogenic CO_2 driving the Earth's climate."

—**Elliott D. Bloom**, Professor Emeritus, Kavli Institute for Particle Astrophysics and Cosmology, Stanford Linear Accelerator Laboratory (SLAC); Fellow, American Physical Society; Member of the SLAC team with Jerome I. Friedman, Henry W. Kendall and Richard E. Taylor who received the 1990 Nobel Prize in Physics

"The third and substantially expanded edition of the book, *Hot Talk, Cold Science* by Fred Singer (with David Legates and Anthony Lupo), is one of the most important contributions undermining the economically and politically problematic and highly controversial scientific doctrine of man-made global warming. No one has done so much as Professor Singer in this respect. We all are his followers as he has shown how science has been compromised and misused for decades to justify politically and ideologically motivated, disastrous economic policies. The contradiction between *Hot Talk* and *Cold Science* is even more pronounced now than when this marvelous book was first published."

—**Václav Klaus**, former President, Czech Republic; 1st Prime Minister, Czech Republic

"Science thrives on civil and dispassionate debate. It demands the use of our reason; in fact, it is useless without it. Drs. Singer, Legates, and Lupo bring science and reason to a debate that has increasingly been driven by panic and politics. Twenty years have passed since the Second Edition of *Hot Talk, Cold Science*, and while its science has withstood the test of time, the rhetoric surrounding climate change has only grown more alarmist. Now more than ever, the public deserves *Hot Talk, Cold Science*'s thorough scientific and economic analysis of the realities of our environment."

—**Thomas M. McClintock**, U.S. Congressman; Member, House Natural Resources Committee and House Judiciary Committee

"Singer, Legates, and Lupo's book *Hot Talk, Cold Science* is excellent. I remain very concerned that the bulk of the environmental movement continues to ignore so many solvable issues while engendering global hyper-anxiety, particularly in the young generation, over a climate crisis that does not exist. This book presents an understandable and balanced review of the science that will leave readers with two conclusions; that our use of fossil fuels is remarkably benign to the environment and essential to the struggling global underclass, and that we have been greatly misled by those entrusted for sound environmental policy."

—**Ian D. Clark**, Professor of Earth and Environmental Sciences, University of Ottawa, Canada

"*Hot Talk, Cold Science* by the late Dr. Fred Singer, with Drs. David Legates and Anthony Lupo, is an enormously important contribution to the 'unfinished debate on global warming' and written by three highly authoritative and trustworthy climate scientists. I very enthusiastically recommend the book to everyone who wishes to learn why this debate is anything but 'settled.'"

—**Larry S. Bell**, Endowed Professor, Sasakawa International Center for Space Architecture, University of Houston; author, *Climate and Corruption: Politics and Power Behind the Global Warming Hoax* and *Scared Witless: Prophets and Profits of Climate Doom*

"The unsurprising political opposition to climate policy—substantial increases in energy costs combined with trivial future climate impacts—has led its proponents to make 'crisis' assertions as loud as they are unsupported by the evidence. That is why we will never observe in the crisis literature a discussion of 'What We Know We Don't Know.' But the book *Hot Talk, Cold Science* offers that, and more broadly, a serious, objective, and highly informative discussion of actual climate science, evidence, and attendant implications. I recommend this book strongly to anyone in pursuit of an honest discussion of climate phenomena."

—**Benjamin Zycher**, Resident Scholar, American Enterprise Institute

"The great Fred Singer lives on. *Hot Talk, Cold Science* is the best exposition of Fred, with help from David Legates, to engage everyone who still cares to listen, on the twists and turns of climate science. Knowing how selfless Fred has been for the past 25 years, I think this is Fred's quiet way to pay final tribute to those very many of us who believe that science must come first before playing a secondary role to public bullying and political blackmailing. This is the reason why this book is so special and should be on everyone's bookshelf."

—**Wei-Hock ("Willie") Soon**, astrophysicist and geoscientist, Solar and Stellar Physics Division, Harvard-Smithsonian Center for Astrophysics

"Anyone having an opinion or interest in climate change simply has to read the updated third edition of the book, *Hot Talk, Cold Science*. It is a fascinating and remarkable voyage over all the central aspects. No one can read it without learning a lot. It turns and twists the various arguments in a most constructive way. A well of knowledge!"

—**Nils-Axel Mörner**, Emeritus Professor of Paleogeophysics and Geodynamics, Stockholm University, Sweden

"The new, third addition of *Hot Talk, Cold Science* is the capstone of Fred Singer's remarkably lucid and straightforward synthesis of anthropogenerated climate change. It is current, comprehensive and clear—just like Fred—a brilliant and fitting coda to the career of a pioneer."

—**Patrick J. Michaels**, Senior Fellow, Competitive Enterprise Institute; former Research Professor of Environmental Sciences, University of Virginia; former President, American Association of State Climatologists; former Virginia State Climatologist

"There is a battle going on. It is not about the environment. It is not about global warming. It is about power and wealth. In *Hot Talk, Cold Science*, Singer, Legates, and Lupo have provided the intellectual ammunition needed to defeat the well-financed fanatical forces with their frightening false narrative. This does not mean honesty will prevail but it means that with the research and arguments available to us by these three disciplined thinkers, the truth will prevail."

—**Dana T. Rohrabacher**, former U.S. Congressman; former Chairman, Space and Aeronautics Subcommittee; Member, Committee on Foreign Affairs and Committee on Science, Space and Technology, U.S. House of Representatives

"Twenty years forward of the 2nd edition arrives the welcome 3rd edition of *Hot Talk, Cold Science: Global Warming's Unfinished Debate*. Drs. Singer, Legates, and Lupo uphold the vital method of scientific inquiry, which is the only way to improve the current, blurred description of the physical nature of the human impact on the terrestrial ecosystem. Nonetheless, the scientific questions about global warming are largely inseparable (properly so) from policy discussion. The authors present extensively updated and expanded material on both aspects. Where heat explodes in the debate is exemplified by loud claims that quantitative description of the vastly complex terrestrial ecosystem is 'settled science,' a belief destructive to invaluable scientific inquiry. Should piratical, adrenaline-mongering continue to strangle rational thought that is antithetical to improved scientific results? The authors reject ignorance in favor of cool, scientific clarity in the search for illumination in the ways of the terrestrial ecosystem in order to inform best policy. As Singer, Legates, and Lupo do, let's keep the debate scientifically sound and enlightening."

—**Sallie L. Baliunas**, former Staff Astrophysicist, Harvard-Smithsonian Center for Astrophysics

"As a trained statistician, I have long been puzzled by apocalyptic visions of CO_2-driven 'global warming,' when the data clearly show there has been but modest warming in recent decades. This is why I greatly welcome the new edition of Dr. Singer's excellent *Hot Talk, Cold Science: Global Warming's Unfinished Debate*, which confirms my suspicions that much of the 'global warming' agenda is bedeviled by fake news and driven by vested interests. Commitments to 'decarbonize' economies (not least of all in the U.K. and U.S.), at great economic cost, are therefore sadly misguided—especially when China, for example, has absolutely no intention of following suit."

    —**Ruth J. Lea**, C.B.E., Ph.D., Co-Founder, Global Vision; Economic Advisor, Arbuthnot Banking Group; former Chief Economist at Mitsubishi Bank; former Head, Policy Unit, Institute of Directors

"Climate alarmism is a hydra. No matter how often the politics of catastrophe, conformity and compliance are refuted by empirical data, new waves of hysteria emerge to sustain the self-fulfilling desire by acolytes to censure society needlessly and punitively. Fred Singer was an early and brave voice of reason and this new edition of his classic book *Hot Talk, Cold Science* is a stark reminder of the contrast between politics and science in understanding the climate divide."

    —**L. Graham Smith**, Professor Emeritus of Geography, University of Western Ontario, Canada

"Fred Singer, like the little Dutch Boy who stood with his finger in the dike, fought against a flood of ill-advised policy decisions in the early years of the climate debate. Now he with climatologists David Legates and Anthony Lupo have armed us with the 3rd revised and expanded edition of *Hot Talk, Cold Science*. It is loaded with facts and technical details but it is written in a manner that is accessible to the non-expert. This new edition of *Hot Talk, Cold Science* is a timely contribution in a world where scientifically-illiterate policymakers are pushing immediate implementation of policies that could destroy the economies of developed nations and leave many developing nations in poverty, while unfairly benefiting certain nations, like China, that are willing to exploit the situation."

    —**Peter D. Friedman**, Professor of Mechanical Engineering, University of Massachusetts Dartmouth; Member, American Geophysical Union

"Where would we be without scientists who are skeptical about alarming predictions of doom accompanied by the ritualistic prescriptions of sacrifice offered up by self-appointed scientific elites and their enthusiastic collaborators in governments and the media? The answer is that we would be in a very dark place indeed. In this, the third edition, of Fred Singer's book *Hot Talk, Cold Science*—this time with David Legates and Anthony Lupo—the authors again turn a skeptical eye to the nature of the evidence that is available on the Earth's climate and how and why it changes, and on the effects of warming and of cooling. As with previous environmentalist alarms—think DDT, acid rain, and mercury in fish—the authors find that a properly skeptical assessment of the evidence does not support alarm. Nor does it support the crushingly expensive policies advocated by the alarmists and adopted by governments. For all our sakes, let us hope that people will read this book and help to turn the tide against the unscientific fear-mongering and bankrupting policies of the global-warming alarm movement sooner rather than later."

—**Kesten C. Green**, Senior Research Fellow, Ehrenberg-Bass Institute, University of South Australia; Co-Director, The Forecasting Principles; former Director, International Institute of Forecasters

"Without a doubt climate change continues to be the leading social and scientific topic of our times. In this updated and expanded edition of *Hot Talk, Cold Science*, Fred Singer, David Legates, and Anthony Lupo provide a very accessible, comprehensive and balanced overview of the state of climate science, as well as an important discussion of the propaganda and rhetoric used to influence the general public and politicians in the era of social media."

—**R. Timothy Patterson**, Professor of Geology and Director, Paterson Laboratory, affiliated with the Carleton Climate and Environmental Research Group, Carleton Institute of Environmental Science, Global Water Institute, and Carleton Northern Studies Program; Carleton University, Canada

"If you are ever involved in climate-change debates, we expect that the argument goes much like this: First, you are told that the 'science is settled,' that 97% of scientists agree that a climate crisis is real, and that humans are the main cause. Second, the planet will suffer irreparable harm if something is not done immediately. In *Hot Talk, Cold Science*, Fred Singer (with David R. Legates and Anthony R. Lupo) brings some cold science to this hot debate. They provide a rational, comprehensive, and thoroughly documented discussion that will appeal to the general public as well as those in the scientific community."

—**Randy T Simmons**, Professor of Political Economy, Department of Economics and Finance, Utah State University; author, *Nature Unbound: Bureaucracy vs. the Environment* (with Ryan Yonk and Kenneth Sim)

"Listening to the IPCC on climate change is like blasting your ears with a fire alarm you can't turn off. In contrast, Dr. Singer has been completely consistent over the years with his message that climate change is not a single variable problem; that water vapor, not carbon dioxide, is far more important; and that natural variables such as the oceans, the clouds and solar activity have been simply left out by mainstream climate science. Now with his superb book *Hot Talk, Cold Science*, it is way past time we listened to his message."

—**Curt G. Rose**, Emeritus Professor of Environmental Studies and Geography, Bishop's University, Canada

"In *Hot Talk, Cold Science*, eminent climatologists Fred Singer, David Legates, and Anthony Lupo reveal many flaws that underlie modern global warming theory. Most importantly, perhaps, is their exploration of scientific modeling—necessary at times, but unreliable for important policy decisions. As they show, such models can be and have been manipulated to advance beliefs and opinions under the guise of science. This book makes an indispensable contribution to a crucial debate."

> —**Ronald J. Rychlak**, Distinguished Professor of Law and Jamie L. Whitten Chair of Law & Government, University of Mississippi School of Law

"*Hot Talk, Cold Science* is readable to the layperson and nicely lays out the arguments. It is a thorough collection of the main climate skeptical arguments—backed by science, not just hype. It answers the questions: 'Why do some very smart people still not accept such arguments? How can they possibly still be deluded?' Read this book."

> —**Peter V. Bias**, William F. Chatlos Professor of Business and Economics, Florida Southern College

"I have read many of Fred Singer's scientific publications, including his NIPCC reports and essays—always courageous and unbiased. To read his book *Hot Talk, Cold Science* for the first time is an eye-opener to learn how Singer at an early stage discovered how a sound scientific debate became mixed with politics and how this has created a distrust of science. May this important book lead us back to an open scientific debate without political pressure. The complex science of climate change has to be investigated from many angles and Fred Singer has shown us how."

> —**Jan-Erik Solheim**, Professor Emeritus, Institute of Theoretical Astrophysics, University of Oslo, Norway

"*Hot Talk, Cold Science* is surely the most complete and insightful book on the convoluted subject of climate change—both from a historical scientific perspective and a political game-like one. Fantastic indeed, as a reference, and also posing the right questions on this very complex and important subject."

> —**Peter Stilb**, Emeritus Professor of Chemistry, KTH, Royal Institute of Technology, Sweden

"The products of science constitute one of the most strikingly glorious examples of the reasoning mind in action, producing the advanced civilization that many enjoy today. But, through the push of so-called 'climate change/global warming' alarmists, science has been under attack by those who capitalize on its name while destroying its methods. The attack is global and has come through scientific organizations, national academies, interest groups, the media, and politicians. In order for ordinary citizens to understand the attack, they would immeasurably benefit by the clear explanations (supported by extensive references) in *Hot Talk, Cold Science* by S. Fred Singer, a scientist of high integrity and honesty who has made an extensive study of the alarmist claims over many years. I think that *Hot Talk, Cold Science* would greatly help people to understand (using a related quote from the book) the 'terrible crime against science, the adoption of unnecessary and very costly policies, and grave damage to the reputation and credibility of science.'"

—**Laurence I. Gould**, Professor of Physics, University of Hartford; former Chair, New England Section of the American Physical Society

"We have all repeatedly heard about global warming and the alleged future disasters caused by our way of life, by media, politicians and scientists who paint a disastrous future due to man-made climate change. Unfortunately, this is a subject with significant political and economic interests that are well known to twist what is right and wrong. With the book, *Hot Talk, Cold Science* by Drs. S. Fred Singer, David R. Legates, and Anthony R. Lupo, we have the opportunity to receive a second opinion that usually cannot be heard on the science and politics. Read it, and you will be better-informed."

—**Henrik Svensmark**, Professor, Division of Solar System Physics, Danish National Space Institute, Technical University of Denmark

"In *Hot Talk, Cold Science*, Fred Singer, David Legates, and Anthony Lupo make a strong case that politics, not sound science, is in the driver's seat of the fears of dangerous, manmade climate change. They provide overwhelming scientific evidence that climate change, though having always been real, is not overwhelmingly manmade and is not even one of the most serious problems facing mankind—let alone an existential threat. They also show precisely what political agendas lie behind the alarmist message—and what must be done to counter them before they undermine freedom and human well-being throughout the world, trapping billions in bondage and poverty. *Hot Talk, Cold Science* is brilliant, comprehensive, timely, and thoroughly vetted."

—**E. Calvin Beisner**, Ph.D., Founder, Cornwall Alliance for the Stewardship of Creation

"*Hot Talk, Cold Science* is very well-written and quite informative for the climate science community at large. The book also provides excellent examples of the flawed science of anthropogenic warming (AGW) and the meaningless push by the advocates of AGW and also by climate modelers to 'Reduce CO_2,' which will be enormously expensive and have minuscule impact on the Earth's climate."

—**Madhav L. Khandekar**, former Research Scientist, Environment Canada; Expert Reviewer, IPCC 2007 Climate Change Documents; Member, Editorial Board, *Journal of Natural Hazards*

"It takes real science to get rockets into space and design the instruments that collect environmental data. Yet many of the people and agencies who have handled this data have altered them by 'administrative adjustments,' using methods that are not revealed, and then use the output in computer simulations designed to justify alarmist hypotheses about future weather and climate. This is definitely not the scientific method. Fred Singer has designed rockets and their instruments, and he has also wrestled with the challenges of data interpretation and misinterpretation. With his unique qualifications, his book *Hot Talk, Cold Science* tells the story of global warming, exposing the rotten core of pseudoscience, the corrupt politics and the dirty tricks. He has done it beautifully."

> —**Clifford D. Ollier**, Professor Emeritus of Geology and Research Fellow, School of Earth and Geographical Sciences, University of Western Australia

"In *Hot Talk, Cold Science*, Fred Singer has elegantly summarized the almost complete dichotomy between the actual science indications that greenhouse gases have a very modest effect on climate at best, and the alarmist view promoted by U.N., I.P.C.C. Science and embraced by most governments worldwide. Derived policies will result in huge expenditure, little discernible effect on climate and negative effects on the standard of living. Unfortunately, all governments financially support almost all science and education, meaning there has now been more than 25 years of sponsored indoctrinated education of our young people worldwide. It seems therefore unlikely that climate policy will change any time soon. The likelihood of any real-world progress towards U.N., I.P.C.C. or government aims for meaningful emissions reductions is indicated by the fact that the world's emissions have increased by about 65% since 1990, a period of time during which most world country governments have professed to embrace policies to achieve exactly the opposite. Eventually sanity will prevail perhaps catalyzed by increasing unrest from various social, industry and political groups throughout the world bringing real policy change. We have already seen such unrest with the Mouvement des Gilets jaunes in France and agricultural industry protests in Europe as governments try to increase the cost of energy or force draconian reductions in animal agriculture. *Hot Talk, Cold Science* is a complete, concise, factual and very valuable addition to the climate debate. I recommend it to all, and suggest it should be compulsory reading for all elected government, regional and local government officials and policymakers."

> —**Arthur John Allison**, ONZM, Fellow, New Zealand Institute of Primary Industry Management; former Director of Agricultural Research for the Southern South Island, Ministry of Agriculture and Forestry, New Zealand; former Manager Director, Silverstream Ltd.; Member, New Zealand Climate Science Coalition

"A real and open debate about global warming and its policy implications has not yet begun. *Hot Talk, Cold Science* challenges conventional wisdom and should be read with an open mind."

> —**Benny J. Peiser**, former Senior Lecturer, Faculty of Science, Liverpool John Moores University; Director, Global Warming Policy Foundation; Founder, Cambridge Conference Network

Praise for Second Edition of *Hot Talk, Cold Science*

"In *Hot Talk, Cold Science*, the illustrious Fred Singer dares to point out that 'the Emperor has no clothes.' Is there evidence to suggest 'discernible human influence' on global climate? Of great interest, this book demonstrates that at best, the available evidence is sketchy and incomplete. *Hot Talk, Cold Science* should have widespread circulation."

— **Arthur C. Clarke**, scientist and author, *2001: A Space Odyssey*

"The scientific urge to consensus on the greenhouse issue tends to compromise away dissent. Fred Singer, with impeccable credentials, does not compromise. His criticism is crucial to the current debate—a debate that will not soon be settled."

— **Thomas C. Schelling**, Nobel Laureate in Economic Sciences; Professor of Economics, University of Maryland

"Where to Learn More About Climate Change: In *Hot Talk, Cold Science*, an atmospheric scientist writes that the scientific community is far from a consensus on the causes and repercussions of global warming."

— *The New York Times*

"*Hot Talk, Cold Science* is a unique and in-depth analysis of an important component of the global warming issue. It presents evidence contrary to the hypothesis that global warming is an immediate and serious threat from man-induced greenhouse gas emissions. Singer's organization of observational material is good and supports the argument that any climate change that has occurred over the last century has been natural and not man-induced. I would encourage anyone involved with the global warming issue to give this book a serious hearing. His view that the evidence does not support the industrial nations taking action at this time to pass laws at reducing fossil fuel emissions in the belief that a significant future global temperature reduction would occur is well thought out. More research and debate are needed on the subject before mandatory restrictions are imposed. He is right, this topic is definitely 'unfinished *business*.'"

— **William M. Gray**, Professor of Atmospheric Science, Colorado State University

"*Hot Talk, Cold Science* is a very effective book to finally initiate constructive discussion on this topic."

— **Roger A. Pielke Sr.**, Professor of Atmospheric Science and Senior Research Scientist, Cooperative Institute for Research in Environmental Sciences, Colorado State University

"For those who believe that the collapse of the Kyoto summit would herald an environmental disaster, we suggest an antidote: a book called *Hot Talk, Cold Science: Global Warming's Unfinished Debate*, by S. Fred Singer (Independent Institute, Oakland, California). Singer, a pioneer in the development of weather satellites, demands that we drag the global warming debate back to the fundamental issue: Is global warming taking place? Look at the date, this physicist advises, warning that computer models that predict global warming in the future are unable to verify the present climate. There are simply not enough data to verify global climate change caused by human activity."

—*Barron's*

"*Hot Talk, Cold Science* is an important, comprehensive, timely, and thorough book."

—**William A. Nierenberg**, Director Emeritus, Scripps Institution of Oceanography

"In an important new book, *Hot Talk, Cold Science*, S. Fred Singer sums up the evidence on global warming as 'neither settled, nor compelling, nor even very convincing.'"

—*Chicago Tribune*

"In *Hot Talk, Cold Science*, Singer has made an important scientific contribution to the global warming debate—a debate that some have attempted to quash or declare concluded. Let us depend on, rather than fear. a healthy scientific debate."

—**Frank H. Murkowski**, U.S. Senator; Chairman, Committee on Energy and Natural Resources

"The arguments in *Hot Talk, Cold Science* are not advanced as idle speculation but firmly crafted from reliable data and almost seamless logic. Opting finally for prudent measures that include conservation and efficiency as well as nuclear and alternative energy sources, the author hedges his bets and agrees to a 'no regrets' policy which includes adaptation to climate change—just in case."

—*Choice*

"*Hot Talk, Cold Science* is the outstanding, up-to-date, scholarly and objective analysis of current conflicting viewpoints on all the key issues relating to greenhouse warming. I highly recommend this excellent book to specialists and non-specialists alike who are interested in determining the state of the science and the arguments and data behind the various aspects of this important issue."

—**Hugh W. Ellsaesser**, Participating Scientist, Atmospheric and Geophysical Sciences, E. O. Lawrence Livermore National Laboratory

"In *Hot Talk, Cold Science*, Fred Singer does not accept global warming. … The references and index are both quite thoroughly done. … This a book that deserves to be read and digested as good arguments are made and if nothing else, the book can serve as an effective 'devil's advocate' for those who may think greenhouse warming is real."

—*EOS-Transactions of the American Geophysical Union*

"*Hot Talk, Cold Science* is of much interest: I've long followed the literature on the long and short-term temperature records, ice ages, etc. I like especially the many charts and graphs in the book. One of the refreshing ideas is that the worry about the melting of the icecaps could be the reverse: enough increased evaporation from the oceans to make the caps grow!"

—**H. Richard Crane**, Member, National Academy of Sciences

"The author of *Hot Talk, Cold Science*, S. Fred Singer maintains that the proposals put forth at Kyoto were based on forecasts from flawed computer models of the earth's climate, and not on actual observations, and he has urged policymakers to adopt a 'no regrets' policy of continued research and unimpeded economic growth. In *Hot Talk, Cold Science*, Singer examines the literature on climate change and lays out a case against the likelihood of an imminent, catastrophic global warming. He also cites evidence suggesting that even if global warming were to occur, it would largely be benign and may even improve human well-being."

—*NOIA Washington Report*, National Ocean Industries Association

"*Hot Talk, Cold Science* is an important, comprehensive book on the issues involved in the raging debate over global warming. This book is essential reading for anyone seeking to understand this problem, especially as it affects working Americans."

—**William J. Cunningham**, Economist, AFL-CIO

"This book claims that global warming is greatly exaggerated. … Singer is correct in depicting the fundamental physics of climate change based upon the increase in greenhouse gases such as carbon dioxide. … *Hot Talk, Cold Science* is useful in that it contains in concise form virtually all of the skeptic's views about climate change."

—*Bulletin of the American Meteorological Society*

"The scientist who set up the American weather satellite system, Dr. S. Fred Singer, has expressed great skepticism as to whether the globe has in fact gotten any warmer in recent years. The temperature readings from the weather satellites don't show it. The careful analysis of data from a variety of sources by Dr. Singer in his book, *Hot Talk, Cold Science*, is in sharp contrast to the hysterical simplicities of the 'global warming' zealots and politicians."

—**Thomas Sowell**, Senior Fellow, Hoover Institution, Stanford University

"*Hot Talk, Cold Science* is essential reading for those who wish to be fully informed about global warming. The importance of this book is its review of the scientific, economic and policy background on global warming with a reasoned assessment of these facts."

—*Environmental Geology*

HOT TALK, COLD SCIENCE

Global Warming's Unfinished Debate

Revised and Expanded Third Edition

S. Fred Singer

With **David R. Legates** and **Anthony R. Lupo**

Forewords by
Frederick Seitz and **William Happer**

INDEPENDENT
I N S T I T U T E

OAKLAND, CALIFORNIA

Independent Institute
100 Swan Way, Oakland, CA 94621–1428
Telephone: 510–632–1366
Fax: 510–568–6040
Email: info@independent.org
Website: www.independent.org

Cover Design: Denise Tsui
Cover Image: loops7 / Getty Images

Library of Congress Cataloging-in-Publication Data Available

ISBN: 978-1-59813-341-7

Contents

Acknowledgments

WE ARE MOST grateful to the Independent Institute and its president, David J. Theroux, who first conceived of this book, and whose assistance has been essential throughout in making all three editions of the book possible. This Third Edition benefited significantly from the research and editorial assistance provided by Joseph and Diane Bast; as well as subsequent review by John R. Christy, Richard S. Lindzen, and Thomas P. Sheahen. Many other colleagues associated with the Science and Environmental Policy Project (SEPP) and the Nongovernmental International Panel on Climate Change (NIPCC) critically reviewed the manuscript and made important scientific contributions to this book.

We gratefully acknowledge the research support received from numerous sources including the Atlas Network, Electric Power Research Institute, Lynde and Harry Bradley Foundation, Smith-Richardson Foundation, Jacobs Family Foundation, Heartland Institute, and Independent Institute. Research assistance for early editions was provided by Sean R. McDonald, editorial assistance by Candace C. Crandall, and general support by Douglas P. Houts. Editorial and publication assistance for the Third Edition was provided by Christopher B. Briggs, George L. Tibbitts, and Jean Blomquist.

Climate change is a complicated subject, touching on many disciplines. The Independent Institute would be grateful for critical comments from readers, and every effort will be made to incorporate these in future printings.

Foreword to First and Second Editions

FOR SCIENTISTS WANTING fame and fortune, it has become far easier to pander to irrational fears of environmental calamity than to challenge them. But Professor Fred Singer has never been one to lean on conventional wisdom. An atmospheric and space physicist, he has unassailable scientific credentials. This book, *Hot Talk, Cold Science*, will be difficult to dismiss, though many, in their rush to establish international agreements and poorly conceived policies and regulations, will undoubtedly wish to do so.

Fred Singer has been a pioneer in many ways. As an academic scientist in the 1950s, he published the first studies on subatomic particles trapped in the Earth's magnetic field—radiation belts later discovered by physicist James Van Allen. Also, in challenging the findings of other scientists, he was the first to make the correct calculations for using atomic clocks in orbit, hence contributing to the verification by satellites of Einstein's general theory. He further designed satellites and instrumentation for remote sensing of the atmosphere, accomplishments for which he received a White House Presidential Commendation.

Switching careers in the 1960s, he established and served as first director of the US Weather Satellite Service, now part of the National Oceanographic and Atmospheric Administration (NOAA); his efforts were recognized with the US Department of Commerce Gold Medal Award. Dr. Robert M. White, former NOAA administrator and later president of the National Academy of Engineering, wrote of Singer's achievement: "The contribution that Fred made to the development of the operational weather satellite system was crucial to its successful launch . . . . His understanding of space technology and remote sensing put him in an outstanding position to chart the course of that very important component . . . some of his fundamental ideas about the use of space vehicles for atmospheric observation were turned into reality."

Returning to university life in the 1970s, Fred Singer's concern with the environment led him to investigate the effects of human activities on the atmosphere. In 1971, he calculated that population growth (together with increased rice growing and cattle raising) would cause a substantial upward trend of methane, an important greenhouse gas that could contribute to climate warming. He also predicted that methane, once it reached the stratosphere, would be transformed into water vapor, leading to a possible depletion of stratospheric ozone. The fact that methane levels are indeed rising was discovered a few years later; that stratospheric water vapor is also increasing was finally demonstrated in 1995.

At the core of Fred Singer's arguments on the global warming issue is a desire to more fully understand the mechanisms that cause climate to change—in response to natural or man-made forcing—and, perhaps more important, to secure a place for science outside the realm of selfish bureaucracy or the reach of irrational environmentalism.

It is one thing to impose drastic measures and harsh economic penalties when an environmental problem is clear-cut and severe. It is quite another to do so when the environmental problem is largely hypothetical and not substantiated by careful observations. This is definitely the case with global warming. As Professor Singer demonstrates—and his views are backed by many in the scientific profession, including myself—we do not at present have convincing evidence of any significant climate change from other than natural causes.

Until we do, it would be a reckless breach of trust to put in force hasty policies that create real personal and economic hardships for most of the world's population.

FREDERICK SEITZ
President Emeritus, Rockefeller University
Past President, National Academy of Sciences

Foreword to Third Edition

DR. S. FRED SINGER, the author of *Hot Talk, Cold Science: Global Warming's Unfinished Debate*, came to the United States from his native Austria as a teenager, a refugee from the Nazis. Soon after the end of World War II, Dr. Singer wrote his physics PhD dissertation on the topic of cosmic rays in Earth's atmosphere. His thesis adviser at Princeton University, Professor John Wheeler, was a pioneer in nuclear physics who had worked with Nobel Prize Laureate Niels Bohr. Dr. Singer's *Doktor Großvater* invented the quantum mechanics of atoms and molecules, a marvelously successful theory that continues to play an important role in climate science. As this book demonstrates, Dr. Singer never lost his youthful fascination with the properties of Earth's atmosphere.

Professor Frederick Seitz's foreword to the First and Second Editions remains true for this Third Edition. With each successive edition, evidence of any harm from increasing levels of carbon dioxide from human activities has become less persuasive. Warming has been much smaller than model predictions. There is still much Arctic ice, even in midsummer. Sea levels are rising at the same rates as a century ago. Extreme weather has not become more frequent or severe.

In spite of the failures of alarmist predictions, or perhaps because of those failures, virtue-signaling elites of the privileged world have become increasingly hysterical about the supposed threat from the "climate emergency" or "climate crisis." The shrill demands of "Green New Dealers" are reminiscent of the "academy of projectors in Lagado" in Jonathan Swift's *Gulliver's Travels*. They had brilliant ideas on how to create an ideal world, which they would manage. But after forty years of effort, "none of these projects are yet brought to perfection; and in the meantime, the whole country lies miserably waste, the houses in ruins, and the people without food or clothes. By all which, instead of being discouraged, they are fifty times more violently bent upon prosecuting their schemes, driven equally on by hope and despair."

The Third Edition of *Hot Talk, Cold Science* contains much new material, ranging from detailed science to discussions of the very nonscientific forces that have driven this latest popular delusion of humanity. Although packed with quantitative detail, the book is aimed at intelligent readers with little background in the hard sciences. As a result, it is an excellent primer for newcomers to this fractious subject. The book will also be useful to those who have spent years immersed in climate science. And, there are many references to original research papers and books for those inclined to dig more deeply.

Dr. Singer discusses not only widely supported theories, but also plausible but more speculative views, not all of which can be right. But the stone that builders reject can become the cornerstone of a scientific discipline. Restricting the discussion to one popular theory or another is quite likely to miss the ultimate truth in a young field like climate science. Earth's climate certainly has future surprises in store for us.

Hot Talk, Cold Science is now a classic. The Third Edition would be a very useful addition to the personal library of those who are seriously interested in cold science as opposed to hot talk.

WILLIAM HAPPER
Cyrus Fogg Bracket Professor of Physics Emeritus, Princeton University

Preface to Third Edition

IT IS NOW twenty-one years after the Second Edition of 1999, which expanded the First Edition of 1997. This is the perfect time to publish this Third Edition; recent developments in politics, climate science, and economics signal the end of concern over global warming (now called "climate change") by the general public.

Policies to control climate change by limiting global carbon dioxide (CO_2) emissions reached their apex with the 1997 Kyoto Protocol, a treaty that extended the 1992 United Nations Framework Convention on Climate Change (FCCC). The Kyoto Protocol never succeeded in reducing global CO_2 emissions; instead, their concentration in the atmosphere has increased about 1 to 2 parts per million per year since 1997.

There have been attempts to make the Kyoto Protocol more effective, notably during a Conference of the Parties (COP) in Copenhagen in 2009 that failed to produce an agreement and a COP in Paris in 2015 that produced the "Paris Agreement," a nontreaty establishing voluntary emission goals for each nation and no penalties. US President Barack Obama never submitted the accord to the US Senate for approval, and President Donald J. Trump announced in 2017 that the United States would cease all participation in the accord and would formally withdraw as quickly as the agreement allowed. Few of the countries that signed onto the Paris Agreement are meeting their voluntary emission reduction targets.

The buildup to both the Copenhagen and Paris COPs was tremendous, with lurid predictions of temperature catastrophes and world-wide inundations of coastal cities. The IPCC, created by the United Nations to bang the drum for a binding international treaty, has produced a series of reports declaring its utter certainty that anthropogenic global warming (AGW) is a crisis, but it has lost the scientific debate. As explained in this new edition, CO_2 has not caused temperatures to rise or weather to worsen or sea level rise to accelerate beyond historical rates. Most scientists now believe the climate is less sensitive to CO_2 than claimed by the IPCC.

Finally, the economics of climate change argues against efforts to dramatically reduce CO_2 emissions (primarily by reducing the use of fossil fuels), favoring instead a "no regrets" policy of reducing emissions only if doing so is justified by other benefits or costs less than what the likely benefits are worth. Developing countries realize their economic development is jeopardized by limiting energy generation and electricity production, and so refuse to make commitments to reduce their emissions. Even the IPCC, in its latest (fifth) Assessment Report, admits the cost of reducing emissions enough to prevent warming would outweigh the benefits.

This new edition has been reorganized, with Part 1 focusing on "hot talk"—the rhetoric and propaganda used to advance the political objective of restricting access to affordable energy—and Part 2 focusing on "cold science"—the physical science and also the cost-benefit analysis and economics revealing that AGW is not a crisis after all. Many of the sources cited in the Second Edition have been replaced by or supplemented with newer sources, and I have tapped some of my writing published in other venues over the past two decades to bring the analysis up-to-date.

Once again, I invite readers who see errors or omissions in this book to inform the Independent Institute so future editions can be improved.

S. FRED SINGER
Rockville, Maryland

List of Figures and Boxes

List of Figures

List of Boxes

PART ONE

Hot Talk

I

A Century of Climate Concerns

CONCERN THAT A man-made increase in greenhouse gases (GHGs) could cause global warming goes back to the early nineteenth century. The first definitive paper calculating a rise in temperature was published by the Swedish chemist Svante Arrhenius in 1896. Even earlier, in 1824, the French mathematical physicist Joseph Fourier showed that certain minor gases, like carbon dioxide (CO_2), in the atmosphere could absorb infrared (IR) radiation (heat) emanating from the Earth's surface, interfere with its escape, and thus raise the temperature of the surface appreciably. He compared the effect to that of a greenhouse.

An informative article about the early thinking on greenhouse warming (Weart 1997) reviews the publications that put the subject on the map. In 1938, the British steam engineer G. S. Callendar revived the largely ignored greenhouse hypothesis of Arrhenius, asserting that the temperature rise since the 1890s was due to greenhouse warming by CO_2. His views were dismissed by his contemporaries. A prominent textbook by T. A. Blair, *Climatology: General and Regional* (Prentice-Hall, 1942, 118), states that "we can say with confidence that climate ... is not influenced by the activities of man, except locally and transiently."

Little Early Support for the Greenhouse Theory

As late as 1955, experts argued that the ocean would take up all of the CO_2 entering the atmosphere from fossil-fuel burning, thus invalidating Callendar's arguments. A more influential objection was that water vapor would dominate greenhouse warming by covering the same spectral region as the CO_2 absorption bands; therefore, CO_2 would not add to the greenhouse effect. The American Meteorological Society's 1951 *Compendium of Meteorology*, edited by Thomas F. Malone, stated that the CO_2 theory was never widely accepted and "was abandoned when it was found that all of the long-wave radiation [that might be] absorbed by CO_2 is [already] absorbed by water vapor" (Malone 1951, 1016).

In 1956, physicist Gilbert Plass, at Johns Hopkins University, pointed out that the CO_2 absorption lines did not coincide with water vapor lines once the pressure-broadening effect of the lower troposphere was removed. The *Compendium* may have been correct, however, in stating that the recent (before 1945) global temperature rise was not related to human activities, but of natural origin.

While it has been known for decades that the atmospheric concentration of CO_2 was rising because of the burning of fossil fuels—oil, gas, and especially coal—precise measurements on a global scale were started only during the International Geophysical Year in 1957 by Charles David Keeling and Roger Revelle. Their data showed large seasonal fluctuations but also a steady upward trend in the level of CO_2. (See Figure 1.) Other researchers quickly observed that the "Keeling curve" seemed to match preindustrial concentrations of CO_2, as measured by ice cores. (See Figure 2.) Note, however, that it is bad practice to join sets of disparate measurements—taken by different instruments at different locations with different levels of accuracy.

At about the same time, Hans Suess at the Scripps Institution of Oceanography became interested in the ocean uptake of CO_2 (Revelle and Suess 1957). But it was Roger Revelle, then director of the Scripps Institution of Oceanography, who was motivated to start long-term measurements of atmospheric CO_2 and, together with Charles David Keeling, discovered that about half of the CO_2 from fossil-fuel burning was remaining in the atmosphere. Revelle, the "father of greenhouse warming," regarded the CO_2 increase as a "grand geophysical experiment" that would reveal the consequences of human intervention with the atmosphere.

In 1971, S. Ichtiaque Rasool and Stephen H. Schneider, two scientists with the National Aeronautics and Space Administration (NASA), estimated an equilibrium climate sensitivity (ECS, the amount of warming expected to result from a doubling of atmospheric CO2 as the climate system tends toward equilibrium [>1,000 years]) of 0.6°C upped to 0.8°C with a simple water vapor feedback.

In a paper delivered at a 1975 conference of the American Academy of Arts and Sciences, Revelle pointed to the beneficial effects of CO2 on agriculture and speculated that the improvements in yield of this century might be connected to a CO2 rise (Revelle 1977). In 1979, the Ad Hoc Study Group on Carbon Dioxide and Climate was convened by the National Academy of Sciences and chaired by Jule G. Charney, an MIT meteorologist (Charney et al. 1979). Those authors conceded the increase in temperatures from a doubling of atmospheric CO2 would be modest, probably not measurable at that time. However, they speculated that with water vapor feedback, a doubling of CO2 would increase atmospheric temperatures sufficiently to result in an increase of surface temperatures by 1.5° to 4.5°C.

Warming or Cooling?

One reason for the lack of concern before the 1980s was the fact that air temperatures in the Northern Hemisphere had been dropping steadily since about 1940. In the early 1970s, an increasing number of climate scientists as well as popularizers were becoming concerned about this downward trend; those of a catastrophic bent saw the temperature decrease as an indicator of a returning ice age.

The literature of the time is filled with technical and popular papers expounding on the possibility of man-made global cooling. A good example is Stephen Schneider's book *The Genesis Strategy: Climate and Global Survival* (1976). Others, like Reid Bryson of the University of Wisconsin and J. Murray Mitchell, a National Oceanic and Atmospheric Administration (NOAA) climatologist, both contributors to *Global Effects of Environmental Pollution* (1970), ascribed the cooling trend to increasing amounts of air pollution and the increased albedo produced by aerosols, mainly from sulfur oxides emitted in coal burning. Their work was largely ignored at the time.

Aerosols were "rediscovered" after the publication of the first IPCC assessments in 1990 and 1992. By 1995, as climate model results and observations were seen to diverge markedly, aerosols were reintroduced as a means

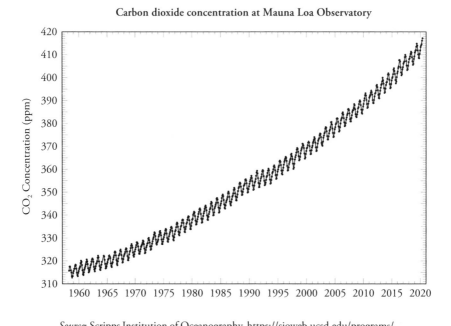

Carbon dioxide concentration at Mauna Loa Observatory

Source: Scripps Institution of Oceanography, https://sioweb.ucsd.edu/programs/ keelingcurve/.

Figure 1. The Keeling curve, tracking the concentration of CO_2 as measured at the Mauna Loa Observatory in Hawaii. The sawtooth pattern illustrates seasonal variability.

for explaining the discrepancy and for "saving" the large climate sensitivities calculated by conventional general circulation models (GCMs).

In the real world, the climate situation changed around 1975, but this was only realized a decade later. Temperatures started to rise rapidly until about 1980, after which, according to satellite data, temperatures stabilized at about the 1980 level. We still do not know for certain what happened in the mid-1970s; many suspect that a sudden climate transition took place in 1976–77 involving ocean circulation. Hurrell and van Loon (1994) indicate a sudden change in Southern Hemisphere circulation, leading also to a strengthening of the Antarctic vortex. Miller and colleagues (1994, 21) wrote, "During the 1976–1977 winter season, the atmosphere-ocean climate system over the North Pacific Ocean was observed to shift its basic state abruptly," creating

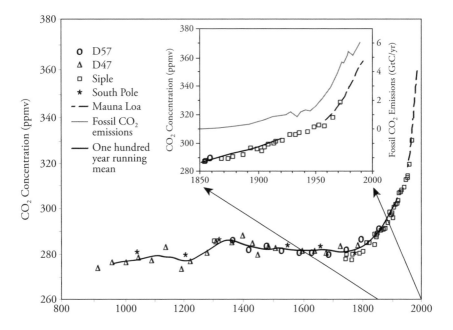

Source: Adapted from IPCC AR2, WGI, 1996.

Figure 2. Preindustrial and recent atmospheric CO_2 concentration as measured in ice cores and at Mauna Loa, Hawaii. Also shown is the estimated emission from the burning of fossil fuels.

a step-function increase in global temperatures. This rather abrupt change is often referred to as the "Great Pacific Climate Shift."

Fear of Global Warming

By 1980, fear of global cooling was replaced by a fear of global warming, overlooking the fact that warming is generally better for agriculture and most other human activities than is cooling. Budding environmental movements were confronted with the fact that air pollution and water pollution levels were decreasing in Western nations, and new problems had to be found to maintain the enthusiasm and funding for the growing organizations. Global disasters filled the bill, and so we had successive scares on acid rain, ozone depletion, and now on greenhouse warming (see Booker 1998, 2009; Darwall 2014). The rise of the Malthusians is described in more detail in Chapter 12.

In a 1984 interview with *Omni* magazine, Roger Revelle expressed optimism and no great alarm about the risk of climate change. But in 1988, a hot summer and major drought devastated much of the agricultural harvest in the United States. In testimony before a Senate committee chaired by then-Senator Albert Gore, NASA climate scientist James Hansen announced he was "99 percent" sure that climate change was here. In letters written in July 1988 replying to climate concerns expressed by his congressman and then-Senator Timothy Wirth, Revelle advised against drawing any conclusions about global warming from the 1988 drought and warned against taking hasty actions.

Hansen's testimony also gave rise to my article on CO_2 published in the *Wall Street Journal* that year titled "Facts and Fancy." In it, I acknowledged that the concentration of several GHGs was increasing because of human activities; that these molecules, because of their inherent radiative properties, enhance the normal greenhouse effect of the atmosphere that relies mainly on existing water vapor, clouds, and CO_2; and that the enhanced greenhouse effect should increase the Earth's average temperature—provided that all other factors remain the same. But the crucial issue, then as now, was whether in fact "other factors remain the same." I wrote, "[A] cottage industry has sprung up on 'climate policy'—not climate science—populated by professional regulators, environmental activists and assorted scientists— all heavily supported by foundations. They attend delightful international conferences, write repetitive papers and testify before important congressional committees—all about a problem that may or may not be real—and which in any case may defy any easy solution." The entire article is reprinted in Box 1 near the end of this chapter.

Shortly before Revelle's death in July 1991, I coauthored an article with him and Chauncey Starr, founder and president of the Electric Power Research Institute (EPRI), published in *Cosmos*. We acknowledged that there are reliable measurements of the increase in GHGs in the earth's atmosphere, but much uncertainty and disagreement about the size of sources and sinks for these gases, whether the increase had caused a change in the climate during the last century, and what future changes to climate would occur as a result of further increases in GHGs. We wrote, "The models used to calculate future climate are not yet good enough because the climate balancing processes are not sufficiently understood, nor are they likely to be good enough until we gain more understanding through observations and experiments. As a consequence, we cannot be sure whether the next century will bring a warming that is negligible or a warming that is significant. Finally, even if there are a global warming and associated climate changes, it is debatable whether the consequences will be good or bad; likely some places on the planet would benefit, some would suffer." This article appears in Box 2 at the end of this chapter. The article caught the attention of Al Gore, who then persuaded a former associate of Revelle named Justin Lancaster to claim that Revelle had not really been a coauthor of the essay. Lancaster also made the ludicrous claim that I had put his name on the paper as a coauthor "over his objections." The claim was completely false and I sued Lancaster for libel. The libel suit was successful. Lancaster signed a retraction and apology. The story is told in a chapter I wrote for a book published by Hoover Press (Singer 2003).

* * *

I believe my *Wall Street Journal* essay and the *Cosmos* essay coauthored with Revelle and Starr accurately reflected the views of most scientists who were studying the global warming issue at the time they were written. However, starting with the scientific meetings at Villach, Austria (1985) and Bellagio, Italy (1987), science took a back seat to politics. At the Earth Summit in Rio de Janeiro, held in 1992, scare tactics and blatant politics replaced science in the climate debate. That story is told in the next two chapters.

Box 1

On June 23, 1988, NASA climatologist James Hansen testified before a US Senate committee that he was "99 percent sure" that "the greenhouse effect has been detected, and it is changing our climate now." About two months later, I countered with a commentary published by the Wall

Street Journal *explaining why such claims are speculative and unreliable. Even at this early date, the main issues and parameters of the debate were clear.*

"Fact and Fancy on Greenhouse Earth"

By S. Fred Singer
Wall Street Journal
August 30, 1988

A hot summer, plus drought in parts of the U.S. has renewed longstanding concerns about the atmospheric greenhouse effect and spawned both doomsday scenarios and legislative proposals to stabilize the climate. As usual, we are dealing with a mixture of fact and fancy. Here are some of the facts:

- The concentration of several minor atmospheric constituents is increasing because of human activities. These trace gases include carbon dioxide, mainly from fossil-fuel burning and cutting down of forests; nitrous oxide, mainly from fertilizers; methane from a variety of natural and human sources; and chlorofluorocarbons (CFCs), synthetic chemicals widely used in refrigeration, air conditioning and plastic-foam manufacture.

- These molecules, because of their inherent radiative properties, enhance the normal greenhouse effect of the atmosphere that relies mainly on existing water vapor and carbon dioxide.

- The enhanced greenhouse effect should increase the Earth's average temperature—provided that all other factors remain the same. Any climatic change has a multitude of consequences; some are beneficial, many are not.

Aside from these facts, all the rest is theory at best, speculation at worst. The crucial issue is to what extent "other factors remain the same." In technical language: Will changes in the atmosphere, ocean or land surface reinforce the climate change (thus causing positive feedback) or will these changes counteract and partly cancel the climate warming (negative feedback)? For example, as oceans warm and more water vapor enters the atmosphere, the greenhouse effect will increase somewhat, but so should cloudiness—which can keep out incoming solar radiation and thereby reduce the warming.

More Research Is Needed

The theory of climate change is not yet good enough to provide a sure answer to the fundamental question: How important is the enhanced greenhouse effect? More research is needed on atmospheric physics and on modeling the atmosphere-ocean system. More observations over the past century can positively disentangle climate fluctuation from long-term trends.

Observed trends do not agree with expectation from greenhouse theory. A large temperature increase of 0.6 degree Celsius, or about 1 degree Fahrenheit, occurred between 1880 and 1940, well before human influences were important. (Despite the growth of heavy industry during that period, the amount of fossil fuels burned for energy was small compared with those burned today.) A temperature decline occurred between 1940 and 1965, followed by a sudden warming of about 0.3 degree Fahrenheit since 1975—too short a period to discern a trend.

We have had more than enough examples of inadequate theories during the past decades:

- In the early 1970s it was believed that a fleet of supersonic transports could destroy the stratospheric ozone layer. Now we suspect that the opposite is true—thanks to better data and theories. In fact SST exhausts are likely to counteract the damaging effects of CFCs on ozone.

- Only a few years ago, it was thought that acid rain could be reduced just by cutting smokestack emissions of sulfur dioxide. Now we recognize nitrogen oxides as a culprit as well; without cutting nitrogen oxides, reduction in sulfur dioxide may not be effective.

- "Nuclear winter" was supposed to freeze the Earth and possibly destroy all human existence. Now we realize that while smoke clouds from fires can darken the sky, the temperature may not fall by much. The theory had neglected the possibility that the smoke cloud may act as a heat blanket, causing its own greenhouse effect. Under some circumstances, a low altitude smoke cloud would even warm the Earth, not cool it.

These examples should induce a certain amount of skepticism and make us somewhat more humble about the ability of theory to predict the future of the atmosphere and of climate.

In the meantime, however, a cottage industry has sprung up on "climate policy"—not climate science—populated by professional regulators, environmental activists and assorted scientists—all heavily supported by foundations. They attend delightful international conferences, write repetitive papers and testify before important congressional committees—all about a problem that may or may not be real—and which in any case may defy any easy solution.

Consider some of the remedies proposed:

- Drastically limiting the emission of carbon dioxide means cutting deeply into global energy use. But limiting economic growth condemns the poor, especially in the Third World, to continued poverty, if not outright starvation.

- Substitutes for fossil fuels, such as hydro, geothermal, solar energy, and wind, are all useful in particular applications but not enough to reverse the growth of atmospheric carbon dioxide. In addition, their wide use would require exorbitant capital investments and could be environmentally damaging. (For example, the energy needs of a three-member household could be met by solar cells covering a whole football field's worth of vegetation.) Curiously, the N-word is only occasionally mentioned—yet nuclear energy is the only realistic, abundant, economic and widely accepted energy source that produces no greenhouse effect and little environmental impact—if properly handled.

- Energy conservation is much to be desired, and there are many unexploited opportunities, to be sure. But there are also great costs involved if carried too far, as indoor air pollution, including radon, in energy-efficient buildings. Realistically speaking, more conservation can only nibble at the carbon dioxide problem, not solve it.

- While we might limit the emission of CFCs, and even carbon dioxide and nitrous oxide, by drastic controls and world-wide regulation, no one has figured out what to do about the growing atmospheric concentration of methane, an important greenhouse gas, contributing about 20 percent of the effect—as against 50 percent for carbon dioxide. Scientific data from the past tell us that methane has been increasing steadily from sources and for reasons we don't fully understand. There is little point in making extreme efforts to control one set of gases while leaving another untouched.

No Palm Trees in New York

But the climate can and does change—and we should be aware of the need to adjust to change. In the last interglacial period, 125,000 years ago, sea level was up 20 feet—all without any human help. What should concern us most is a very rapid change in climate, one to which our economy cannot adjust. Adjustment problems certainly would exist for agricultural soils, which require hundreds or thousands of years for their generation. Climate may indeed change, with or without human interference, but there won't be palm trees in New York, cotton in Toronto, or wheat in Labrador—even by the year 2100.

Congress has heard from a reputable scientist, James E. Hansen of NASA's Goddard Institute, who is "99 percent sure" that the greenhouse effect "is here." Perhaps this means that temperature should rise according to the prediction of standard greenhouse theory. That rise is at least 1 degree Fahrenheit per decade: we won't be able to miss it if it happens. Other reputable but less vocal atmospheric scientists estimate the rise as much less, however.

Public policy about whether to take immediate drastic action thus faces the perennial problem of decision-making with incomplete and conflicting scientific information. We need an analysis that weighs the risk from a delay in instituting far-reaching controls against the possibility of substantially improving the science so that predictions will be more certain.

Box 2

Al Gore famously credited Roger Revelle, a distinguished climate scientist, for being his "mentor" on global warming and alerting him to its possible threat. But I knew Revelle, and he did not share Gore's alarmist views on the subject. In 1992, shortly before Revelle's death, I coauthored with him an essay in Cosmos *that plainly expressed our shared skepticism. This innocuous essay led to a successful libel suit against environmental lawyer Justin Lancaster.*

"What to Do About Greenhouse Warming: Look Before You Leap"
By S. Fred Singer, Roger Revelle, and Chauncey Starr
Cosmos: A Journal of Emerging Issues
Vol. 5, No. 2, Summer 1992

Greenhouse warming has emerged as one of the most complex and controversial environmental and foreign-policy issues of the 1990s. It is an environmental issue because carbon dioxide (CO_2), generated from the prolific burning of oil, gas and coal, is thought to enhance, by trapping heat in the atmosphere, the natural greenhouse effect that has kept the planet warm for billions of years. Some scientists predict drastic climatic changes in the 21st century.

It is a foreign-policy issue because, for a number of reasons, the United States has taken a more cautious approach to dealing with CO_2 emissions than have many industrialized nations. Wide acceptance of the Montreal Protocol, which limits and rolls back the manufacture of chlorofluorocarbons (CFCs) to protect the ozone layer, has encouraged environmental activists at international conferences in the past three years to call for similar controls on CO_2 from fossil-fuel burning.

These activists have expressed disappointment with the White House for not supporting immediate action. But should the United States assume "leadership" in a hastily-conceived campaign that could cripple the global economy, or would it be more prudent to assure first, through scientific research, that the problem is both real and urgent?

We can sum up our conclusions in a simple message: The scientific base for a greenhouse warming is too uncertain to justify drastic action at this time.

There is little risk in delaying policy responses to this century old problem since there is every expectation that scientific understanding will be substantially improved within the next decade. Instead of premature and likely ineffective controls on fuel use that would only slow down but not stop the further growth of CO_2, we may prefer to use the same resources—trillions of dollars, by some estimates—to increase our economic and technological resilience so that we can then apply specific remedies as necessary to reduce climate change or to adapt to it.

That is not to say that prudent steps cannot be taken now; indeed, many kinds of energy conservation and efficiency increases make economic sense even without the threat of greenhouse warming.

The Scientific Base

The scientific base for greenhouse warming (GHW) includes some facts, lots of uncertainty and just plain lack of knowledge requiring more observations, better theories and more extensive calculations. Specifically, there are reliable measurements of the increase in so-called greenhouse gases in the Earth's atmosphere, presumably as a result of human activities. There is uncertainty about the strength of sources and sinks for these gases, i.e., their rates of generation and removal. There is major uncertainty and disagreement about whether this increase has caused a change in the climate during the last century. There is also disagreement in the scientific community about predicted future changes as a result of further increases in greenhouse gases. The models used to calculate future climate are not yet good enough because the climate balancing processes are not sufficiently understood, nor are they likely to be good enough until we gain more understanding through observations and experiments.

As a consequence, we cannot be sure whether the next century will bring a warming that is negligible or a warming that is significant. Finally, even if there are a global warming and associated climate changes, it is debatable whether the consequences will be good or bad; likely some places on the planet would benefit, some would suffer.

Greenhouse Gases (GHG)

It has been common knowledge for about a century that the burning of fossil fuels would increase the normal atmospheric content of carbon dioxide (CO_2), causing an enhancement of the natural greenhouse effect and a possible warming of the global climate. Advances in spectroscopy in the last century produced evidence that CO_2 and other molecules made up of more than two atoms absorb infrared radiation and thereby would impede the escape of such heat radiation from the Earth's surface. In fact, it is the greenhouse effect from naturally occurring CO_2 and water vapor (H_2O) that has warmed the Earth's surface for billions of years; without the natural greenhouse effect ours would be a frozen planet without life.

Precise measurements of the increase in atmospheric CO_2 date to the International Geophysical Year of 1957–58. More recently it has been discovered that other greenhouse gases, i.e., gases that absorb strongly in the infrared, have also been increasing at least partly as a result of human activities. They currently produce a greenhouse effect nearly equal to that of CO_2, and could soon outdistance it.

Methane (CH_4) is produced in large part by sources that relate to population growth; among these are rice paddies, cattle, landfills, forest fires, coal mines and oil field operations. Indeed, methane, now 20 percent of the greenhouse effect but growing twice as fast as CO_2, has more than doubled since pre-industrial times; it would soon become the most important greenhouse gas if CO_2 emissions were to stop.

Nitrous oxide (N_2O) has increased by 10 percent, most likely because of soil bacterial action promoted by the increased use of nitrogen fertilizers.

Ozone (O) from urban air pollution adds about 10 percent to the global greenhouse effect. It may decrease in the U.S. as a result of Clean Air legislation but increase in other parts of the world.

CFCs manufactured for use in refrigeration, air conditioning and industrial processes are making an important contribution but will soon be replaced by less-polluting substances.

Water vapor (H_2O) turns out to be the most effective greenhouse gas by far. It is not man-made, but is assumed to amplify the warming effects of the gases produced by human activities. We don't really know whether H_2O has increased in the atmosphere or whether it will increase in the future although that's what all the model calculations assume. Indeed, predictions of future warming depend not only on the amount but also on the horizontal and especially the vertical distribution of H_2O, and on whether it will be in the atmosphere in the form of a gas or as liquid cloud droplets or as ice particles. The current computer models are not complete enough to test these crucial points.

The Climate Record

The issue now is whether the 25 percent increase of CO_2 in the atmosphere, mainly since World War II, calls for immediate and drastic action to limit and roll back global energy use. Taking account of increases in the other trace gases that produce greenhouse effects, we have already gone halfway to an effective GHG doubling; something that cannot be reversed in our lifetime and, according to the prevailing theory, locked in a temperature increase of about 1.6 degrees Celsius.

But has there been a climate effect caused by the increase of greenhouse gases in the last decades? The data are ambiguous to say the least. Advocates for immediate action profess to see a global warming of about 0.5 degrees C since 1880, and point to record global temperatures in the 1980s and the warmest year on record in 1990. Most atmospheric scientists tend to be cautious, however; they call attention to the fact that the greatest temperature increase occurred before the major rise in greenhouse gas concentration. It was followed by a quarter-century decrease between 1940 and 1965 when concern arose about an approaching ice age! Following a sharp increase during 1975–80, there has been no clear upward trend during the 1980s despite some very warm individual years and record GHG increases. Similarly, global atmospheric (rather than surface) temperatures measured by Tiros weather satellites show no trend in the last decade.

Scientists Kirby Hanson, Thomas Karl and George Maul of the National Oceanic and Atmospheric Administration (NOAA) find no overall warming in the U.S. temperature record, contrary to the global record assembled by James Hansen of the National Aeronautics and Space Administration (NASA). Using a technique that eliminates urban "heat islands" and other local distorting effects, climatologist Thomas Karl and colleagues confirm the temperature rise before 1940, followed, however, by general decline. Reginald Newell and colleagues at the Massachusetts Institute of Technology (MIT) report no substantial change in the global sea-surface temperature in the past century; yet the ocean, because of its much greater heat inertia, should control any atmospheric climate change.

Perhaps most interesting are the NOAA studies that document a relative rise in night temperatures in the U.S. in the last 60 years, while daytime values stayed the same or declined. This is just what one would expect from the increase in atmospheric greenhouse gas concentration. But its consequences, as University of Virginia climatologist Patrick Michaels and others have pointed out, are benign: A longer growing season, fewer frosts, no increase in soil evaporation.

It is therefore fair to say that we haven't seen the huge greenhouse warming, of between 0.7 degrees and 2.5 degrees C, expected from the conventional theories. Why not? This scientific puzzle has many suggested solutions:

- The warming has been "soaked up" by the ocean and will appear after a delay of some decades. Plausible but there is no evidence to support this theory until deep-ocean temperatures are measured on a routine basis, as suggested by Scripps Institution oceanographer Walter Munk. Feasibility tests are currently underway, using a sound source at Heard Island in the South Indian Ocean and a global network of microphones, but data over at least a decade will be needed to provide an answer.

- The warming has been overestimated by the existing models. Meteorologists Hugh Ellsaesser (Livermore National Laboratory) and Richard Lindzen (MIT) propose that the models do not take proper account of tropical convection and thereby over-estimate the amplifying effects of water vapor over this important part of the globe. Other atmospheric scientists suggest that the extent of cloudiness may increase as ocean temperatures try to rise and as evaporation increases. Clouds reflect incoming solar radiation; the resultant cooling could offset much of the greenhouse warming. Most intriguing has been the suggestion by British researchers that sulfates from smokestacks—the precursors of acid rain—may have played a role in producing an increase in bright stratocumulus clouds.

- The warming exists as predicted, but has been hidden by offsetting climate changes caused by volcanoes, solar variations, or other natural causes as yet unspecified such as the cooling from an approaching ice age. (Some, like Robert Balling of Arizona State University, consider the warming before 1940 to be a recovery from the Little Ice Age that prevailed from 1600 to about 1850; if correct, this would imply no net warming at all in the past century due to GHW.)

Each hypothesis has vocal proponents and opponents in the scientific community; but the jury is out until better data become available.

Another Ice Age Coming?

Global temperatures have been declining since the dinosaurs roamed the Earth some 70 million years ago. About 2 million years ago, a new "ice age" began—most probably as a result of the drift of the continents and the buildup of mountains. Since that time, the Earth has seen 17 or more cycles of glaciation, interrupted by short (10,000 to 12,000 years) interglacial or warm periods. We are now in such an interglacial interval, the Holocene, that started 10,800 years ago. The onset of the next glacial cycle cannot be very far away.

It is believed that the length of a glaciation cycle, about 100,000 to 120,000 years, is controlled by small changes in the seasonal and latitudinal distribution of solar energy received as a result of changes in the Earth's orbit and spin axis. While the theory can explain the timing, the detailed mechanism is not well understood, especially the sudden transition from full glacial to interglacial warming. Very likely an ocean-atmosphere interaction is triggered and becomes the direct cause of the transition in climate.

The climate record also reveals evidence for major climatic changes on time scales shorter than those for astronomical cycles. During the past millennium, the Earth experienced a "climate Optimum" around 1100 A.D., when Vikings found Greenland to be green and Vinland (Labrador?) able to support grape growing. The "Little Ice Age" found European glaciers advancing well before 1600 and suddenly retreating starting in 1860. The warming reported in the global temperature record since 1880 may thus simply be the escape from this Little Ice Age rather than our entrance into the human greenhouse.

Mathematical Models

Indeed, there is much to complain about when it comes to predictions of future climate, but there is really no alternative to global climate models. "Models are better than handwriting," claims Stephen Schneider of the National Center for Atmospheric Research (NCAR), but how much better? Half a dozen of these General Circulation Models (GCM) are now running mostly in the United States. Even though they use similar basic atmospheric physics, they give different results. There is general agreement among them that there should be global warming; but, with an effective GHG doubling, the calculated average global increase ranges between 1.5 degrees and 4.5 degrees C! These predicted values were unchanged for many years, then crept up and have recently dropped back to the lower end of the range. Just during 1989 some of the modelers cut their predictions in half as they tried to include clouds and ocean currents in a better way. Further, there is serious disagreement among the models on the regional distribution of this warming and on where the increased precipitation will go.

The models are "tuned" to give the right mean temperature and seasonal temperature variation, but they fall short of modeling other important atmospheric processes, such as the poleward transport of energy via ocean currents and atmosphere from its source in the equatorial region. Nor do they encompass longer scale processes that involve the deep layers of the oceans or the ice and snow in the Earth's cryosphere, nor fine-scale processes that involve convection, cloud formation, boundary layers, or that depend on the Earth's detailed topography.

There are serious disagreements also between model results and the actual experience from the climate record of the past decade, according to Hugh Ellsaesser. Existing models retroactively predict a strong warming of the polar regions and of the tropical upper atmosphere, and less warming in the Southern Hemisphere than the northern—all contrary to observations. Yet there is hope that research, including satellite observations and ocean data, will provide many of the answers within this decade. Faster computers will also allow higher spatial resolution and incorporate the detailed and more complicated interactions that are now neglected.

Impacts of Climate Change

Assume what we regard as the most likely outcome: A modest average warming in the next century well below the normal year-to-year variation and mostly at high latitudes and in the winter. Is this necessarily bad? One should perhaps recall that only a decade ago when climate cooling was a looming issue, economists of the National Academy of Sciences' National Research Council calculated a huge national cost associated with such cooling. More to the point perhaps, actual climate cooling, experienced during the Little Ice Age or in the famous

1816 New England "year without a summer," caused large agricultural losses and even famines.

If cooling is bad, then warming should be good, it would seem—provided the warming is slow enough so that adjustment is easy and relatively cost-free. Even though crop varieties are available that can benefit from higher temperatures with either more or less moisture, the soils themselves may not be able to adjust that quickly. But agriculturalists, like Sherwood Idso of the U.S. Department of Agriculture and Yale professor William Reifsnyder, generally expect that with increased atmospheric CO_2—which is, after all, plant food—plants will grow faster and need less water. The warmer night temperatures suggested by Patrick Michaels, using the data of Thomas Karl, translate to longer growing seasons and fewer frosts. Increased global precipitation should also be beneficial to plant growth.

Keep in mind also that year-to-year changes at any location are far greater and more rapid than what might be expected from greenhouse warming; and nature, crops and people are already adapted to such changes. It is the extreme climate events that cause the great ecological and economic problems: crippling winters, persistent droughts, extreme heat spells, killer hurricanes and the like. But there is no indication from modeling or from actual experience that such extreme events would become more frequent if greenhouse warming becomes appreciable. The exception might be tropical cyclones, which Balling and Randall Cerveney argue would be more frequent but weaker, would cool vast areas of the ocean surface and increase annual rainfall. In sum, climate models predict that global precipitation should increase by 10 to 15 percent, and polar temperatures should warm the most, thus reducing the driving force for severe winter-weather events.

There is finally the question of sea-level rise as glaciers melt and fear of catastrophic flooding. The cryosphere certainly contains enough ice to raise sea level by 100 meters; and, conversely, during recent ice ages, enough ice accumulated to drop sea level 100 meters below the present value. But these are extreme possibilities; tidal-gauge records of the past century suggest that sea level has risen modestly, about 0.3 meters. But the gauges measure only relative sea level, and many of the gauge locations have dropped because of land subsidence. Besides, the test locations are too highly concentrated geographically, mostly on the U.S. East Coast, to permit global conclusions. The situation will improve greatly, however, in the next few years as precise absolute global data become available from a variety of satellite systems.

In the meantime, satellite radar-altimeters have already given a surprising result. As reported by NASA scientist Jay Zwally in *Science*, Greenland ice-sheets are gaining in thickness—a net increase in the ice stored in the cryosphere and an inferred drop in sea level—leading to somewhat uncertain predictions about future sea level. Modeling results suggest little warming of the Antarctic Ocean because the heat is convected to deeper levels. It is clearly important to verify these results by other techniques and also get more direct data on current sea-level changes.

Summarizing the available evidence, we conclude that even if significant warming were to occur in the next century, the net impact to the entire planet may well be beneficial with some regions enjoying improved climate, some encountering worse. This would be even more true if the long anticipated ice age were on its way.

In view of the uncertainties about the degree of warming, and the even greater uncertainty about its possible impact, what should we do? During the time that an expanded research program reduces or eliminates these uncertainties, we can be putting into effect policies and pursue approaches that make sense even if the greenhouse effect did not exist.

Energy Policies

Conserve energy by discouraging wasteful use locally. Conservation can best be achieved by pricing rather than by command-and-control methods. If the price can include the external costs that are avoided by the user and loaded onto someone else, this strengthens the argument for proper pricing. The idea is to have the polluter or the beneficiary pay the cost. An example would be peak-pricing for electric power. Yet another example, appropriate to the greenhouse discussion, is to increase the tax on gasoline to make it a true highway-user fee instead of having most capital and maintenance costs paid by various state taxes, as is done now. Congress has lacked the courage for such a direct approach, preferring instead regulation that is mostly ineffective and produces large indirect costs for the consumer.

Improve efficiency in energy use. Energy efficiency should be attainable without much intervention, provided it pays for itself. A good rule of thumb: If it isn't economic, then it probably wastes energy in the process and we shouldn't be doing it. Over-conservation can waste as much energy as under-conservation. (For example, destroying all older cars would certainly raise the fuel efficiency of the fleet, but replacing these cars would consume more energy in their manufacture.) If energy is properly priced, i.e., not subsidized, the job for government is to remove the institutional and other road blocks:

- Provide information to consumers, especially on life-cycle costs for home heating, lighting, refrigerators and other appliances.

- Encourage but not force the turnover and replacement of older, less efficient (and often more polluting) capital equipment: cars, machinery, power plants. Some existing policies that make new equipment too costly go counter to this goal.

- Stimulate the development of more efficient systems, such as a combined-cycle power plant or a more efficient internal combustion engine.

Use non-fossil-fuel energy sources wherever this makes economic sense. Nuclear power is competitive now, and in many countries is cheaper than fossil-fuel power, yet it is often opposed on environmental grounds. The problems cited against nuclear energy, such as disposal of spent nuclear fuel, are more political and psychological than technical. To address safety concerns, nuclear engineers are focusing on an "inherently" safe reactor. Nuclear energy from fusion rather than from fission may be a longer-term possibility, but the time horizon is uncertain.

Solar energy, and other forms of renewable energy, should also become more competitive as their costs drop and as fossil-fuel prices rise. Solar energy applications are restricted not only by cost; solar energy is both highly variable and very dilute; it takes a football field of solar cells to supply the total energy allocated to the average U.S. household. Wind energy and biomass are other forms of solar energy, competitive in certain applications. Schemes to extract energy from temperature differences in the ocean have been suggested as inexhaustible sources of non-polluting hydrogen fuel, once we solve the daunting technical problems.

Direct Interventions

If greenhouse warming ever becomes a problem, there are a number of proposals for removing CO_2 from the atmosphere. Rebuilding forests is widely talked about, but may not be cost-effective; yet natural expansion of boreal forests—those in high-latitude regions—in a warming climate would sequester atmospheric CO_2. A novel idea, proposed by California oceanographer John Martin, is to fertilize the Antarctic Ocean and let plankton growth do the job of converting CO_2 into bio-material. The limiting trace nutrient may be iron, which could be supplied and dispersed economically.

If all else fails, there is always the possibility of putting "Venetian blind" satellites into Earth orbit to modulate the amount of sunshine reaching the Earth. These satellites could also generate electric power and beam it to the Earth, as originally suggested by space pioneer Peter Glaser of A. D. Little, Cambridge, Mass. Such schemes may sound far-fetched, but so did many other futuristic projects in the past and in the present, like covering the Sahara with solar cells or Australia with trees.

Conclusion

Drastic, precipitous and, especially, unilateral steps to delay the putative greenhouse impacts can cost jobs and prosperity and increase the human costs of global poverty, without being effective. Stringent controls enacted now would be economically devastating—particularly for developing countries for whom reduced energy consumption would mean slower rates of economic growth—without being able to delay greatly the growth of greenhouse gases in the atmosphere.

Yale economist William Nordhaus, one of the few who have been trying to deal quantitatively with the economics of the greenhouse effect, has pointed out that "those who argue for strong measures to slow greenhouse warming have reached their conclusion without any discernible analysis of the costs and benefits. . . . " It would be prudent to complete the ongoing and recently expanded research so that we will know what we are doing before we act. "Look before you leap" may still be good advice.

2

FCCC and the Kyoto Protocol

IN 1988, THE same year James Hansen testified to the US Congress that he was "99 percent" sure that climate change was here, the United Nations' Intergovernmental Panel on Climate Change (IPCC) was created by two United Nations agencies, the World Meteorological Organization (WMO) and the United Nations Environment Programme (UNEP). The IPCC produced its first scientific assessment (AR1) in 1990; it was updated in 1992 in time for the Earth Summit held in Rio de Janeiro.

The Earth Summit

At the Earth Summit in 1992, President George H. W. Bush brought the United States into the United Nations Framework Convention on Climate Change (FCCC), a treaty based on the assumption that the GHG theory was correct and calling on the nations of the world to do exactly what Roger Revelle and I had warned against, to start spending trillions of dollars in a vain effort to prevent climate change. Attempts to establish a "protocol" to limit and roll back emissions of GHGs, especially carbon dioxide (CO_2), followed the treaty. The paradigm was the 1987 Montreal Protocol, which led to a ban on the production of ozone-depleting chemicals such as chlorofluorocarbons (CFCs).

The countries that ratified the FCCC convened as a Conference of the Parties (COP-1) in Berlin in 1995. COP-1 produced the Berlin Mandate, instructing the parties to prepare a protocol for implementing the treaty. The second Conference of the Parties (COP-2), meeting in Geneva in July

1996, accepted as a basis for urgent policy action the IPCC's main conclusion about a "discernible human influence on climate." The assembled statesmen chose to regard the science as "settled" and proceeded to plan for COP-3, where countries would present firm proposals for mandatory controls on the emission of GHGs.

The head of the US delegation, Under Secretary of State for Global Affairs Timothy Wirth, proposed legally binding targets and time frames for emissions of GHGs. He stated, "The science calls upon us to take urgent action." When several delegations did not accept Wirth's proposal, a Ministerial Declaration by the United States and like-minded nations was issued on July 18, 1996, calling for a protocol to control emissions of CO_2—and, in effect, to limit the generation of energy.

The policy proposals of the US State Department were presented to a wide audience in a briefing on January 17, 1997. The proposals envisaged national emission "budgets"—which are really quotas, or a certain fraction of the global emission permitted for a particular year. This rationing scheme for the use of fossil fuels was to be supplemented by an emission-trading scheme that would allow any party to buy unused emission rights from any other party. In principle, such trading results in less social cost than strict rationing without trading, but implementing such a scheme requires deciding upon the emission budget to be assigned to each nation, whether to use population or per capita consumption as a criterion, whether to use current values or some future values for population and per capita consumption, and myriad other policy decisions.

COP-3, held in Kyoto, Japan in December 1997, produced the Kyoto Protocol, which is discussed in some detail in this chapter. Many more COPs have taken place since 1997, with the most recent one at the time of this writing being COP-25 held in Madrid in December 2019. The most significant of those meetings was COP-21 held in Paris from November 30 to December 13, 2015, which produced the Paris Agreement, discussed in Chapter 3.

Framework Convention on Climate Change (FCCC)

The announced objective of the FCCC (in Article 2) is to "achieve stabilization of greenhouse gas concentrations in the atmosphere at a level that would prevent dangerous anthropogenic interference with the climate system." The FCCC further states that "policies and measures to deal with climate change should be cost-effective so as to insure global benefits at the lowest cost" (Article 3.3). The FCCC also calls for an "economic system that would lead to sustainable economic growth and development" (Article 3.5).

The FCCC provided a deeply flawed foundation for the Kyoto Protocol, which expired in 2012, and the Paris Agreement adopted in April 2016. It rests on four suppositions that are questionable or even demonstrably false:

1. It assumes that a global warming signal has been detected in the climate record of the past hundred years, thus validating the computer model predictions of a major future warming.

2. It further assumes that a substantial warming in the future will produce catastrophic consequences, including droughts, floods, storms, a rapid and significant rise in sea level, a collapse of agriculture, and a spread of tropical diseases.

3. It assumes that we can avoid the impending climate catastrophe by reducing GHG emissions to some "safe" level, and that this "safe" level is known and achievable.

4. Finally, it assumes that emission reductions should take place regardless of their costs, that the consequences of climate change are so disastrous that no price is too high to pay to prevent it from occurring.

Regarding assumptions #1 and #2, the "cold science" presented in Part 2 of this book will make clear that these declarative scientific statements are unjustified by scientific data. Even the IPCC admitted in its Fifth Assessment Report (2013) that global temperatures did not increase for fifteen years (1998–2012) and that climate models failed to forecast such a "pause," and hence are not validated. Extensive research on the impacts of climate change reveals none of the trends assumed to be occurring by now if human GHG emissions were not dramatically curtailed (NIPCC 2019, esp. chapters 2 and 8).

Regarding assumption #3, "dangerous," like "safe," is a political and not a scientific concept. There are no data revealing the "best" global temperature, amount of precipitation, or amount of ice in the world. Scientists can attempt to measure past and current values for each of these elements of Earth's climate, and can even attempt to predict future values, but whether those values are good or bad (or don't matter) can be determined only by their impacts on people (or perhaps on animals and plants). A physicist doesn't know how much people value fast and safe transportation, affordable electricity and safe heating and cooling of their homes, and countless other goods and services made possible by their use of fossil fuels. Trying to monetize those impacts is the work of economists, and by their own admissions, economists say they cannot assign values to such things (Ackerman et al. 2009; Pindyck 2013; Weitzman, 2015).

Regarding assumption #4, *what to do* about climate change does not emerge full-grown and clothed, as it were, from the scientific discovery of a problem. Effective and fair public policies are developed by weighing the costs and benefits of different plans against one another and against the option of doing nothing at all. In the climate change debate, the costs imposed by the use of fossil fuels on humanity and the environment, including climate change if the science reveals that to be a genuine cost, must be compared to the benefits produced by the continued use of fossil fuels. If the costs of climate change exceed the benefits of fossil fuels, then efforts to force a transition away from fossil fuels are justified and ought to continue. If, on the other hand, the benefits of fossil fuels are found to exceed the costs of climate change, then the right path forward would be to allow energy consumption to increase without interference by governments.

Cost-benefit analysis (CBA) is an economic tool that is widely used in the private and public sectors to determine if the benefits of an investment or spending on a government program exceed its costs (Singer 1979; Hahn and Tetlock 2008; OMB 2013). It is simply astonishing, and unacceptable, that the FCCC does not acknowledge the need to conduct a CBA on its proposed energy-rationing scheme. Since it does not, it is not surprising at all that the IPCC, tasked with providing scientific advice to the FCCC, would likewise fail to conduct such an analysis.

The Kyoto Protocol

The origin of the Kyoto Protocol and its demise is a thrilling tale, full of heroes and villains, which never has been fully told. I was fortunate, if that is the right word, to have been involved continuously in all aspects of the treaty.

The 1997 Protocol, negotiated in Kyoto, Japan, tried to put teeth into the FCCC. Concluded in December 1997, it would oblige Western nations to reduce GHG emissions by an average of 5.2 percent (from 1990 levels) by the period 2008 to 2012. The protocol was finally signed by the United States in November 1998, over the vociferous opposition of many members of the US Senate. In July 1997, the Senate had passed by a vote of 95–0 the Byrd-Hagel Resolution against a Kyoto-like agreement principally because it did not require emission reductions by some 130 developing nations, including China, India, and Brazil. *The United States never did ratify the protocol—*even during the Clinton-Gore years in the White House. In 2001, President George W. Bush withdrew the United States from the protocol. During the Obama administration (2009–17), the Democrat-controlled Senate refused to consider a cap-and-trade bill to restrict emissions of CO_2 that the House had passed in 2009.

The main actor behind the Kyoto Protocol was the IPCC. Its First Assessment Report of 1990 (AR1) provided the basis for the Earth Summit in Rio and its doctored Second Assessment Report (AR2) of 1996 provided the scientific underpinning for the Kyoto Protocol.

What exactly did the IPCC have to say in 1996, when its printed AR2 became available? Those of us present in Madrid in 1995, when a final draft was approved by the scientists, became aware that crucial language was changed *after* its approval by the authors and *before* it was printed. While this has been hotly denied by the perpetrators, the evidence is quite clear; one only has to compare the two documents. As reported in Chapter 4, Dr. Frederick Seitz, one of America's most distinguished scientists and President Emeritus of Rockefeller University, blew the whistle: "In my more than 60 years as a member of the American scientific community, including service as president of both the National Academy of Sciences and the American Physical Society, I have never witnessed a more disturbing corruption of the peer-review process than the events that led to this IPCC report" (Seitz 1996, 16). He had good reason to be upset, because the phrases that were deleted from the final draft would have removed any cause for action by the FCCC parties.

The Kyoto Protocol was a fraud from day one. Even if it had been punctiliously followed by all of the nations that ratified it, it would have achieved essentially nothing. If one were to accept IPCC figures, the additional warming of 1.39°C predicted to occur by 2050 would be reduced to 1.33°C, a reduction of only 0.06°C (Parry et al. 1998). The difference would have been imperceptible. As I wrote at the time,

> But why bother about science? Cynical politicians have pronounced the science "settled" so they can go ahead and negotiate. Like good lawyers that they are, they simply stipulate the scientific conclusions. No more research needed; the science is "complete," "compelling"—or whatever; you scientists can now go away and let us do our job. And for heaven's sake don't come up with any new scientific facts that could mess up our sandbox and ruin our fun (Singer 1998, A19).

Kyoto was all about politics and money. The terms of the Kyoto Protocol demanded a 5.2 percent overall reduction from the emission levels of 1990 for industrialized nations. The choice of 1990 as the base year, however, favored Europe, Britain, Germany, and Russia at the expense of the United States. Around 1990, Britain switched from primarily coal to natural gas, thus reducing its CO_2 emissions. And at about the same time, the Soviet Union collapsed and Germany took over its eastern part, closing down much of its inefficient coal-fired electricity production.

The most pernicious provisions of the Kyoto Protocol were permits for emissions trading within the European Union and the so-called Clean Development Mechanism (CDM). CDM permitted industries and others to keep emitting CO_2 while buying unused credits from other Kyoto nations or by sponsoring projects in developing nations that would reduce emissions. What a racket this turned out to be. It has made Al Gore a "climate billionaire" who emits CO_2 copiously from his four residences, jet planes and yachts, but then buys "carbon offsets," emission credits *from his own company* set up to trade CO_2 permits (Jean 2008).

The other big money item has been the drive for so-called "clean energy," with its huge subsidies for wind power and solar energy, widely abused in

Europe but especially in the United States, where the subsidies are among the highest. The poster child for clean energy is probably ethanol—a huge sink for government subsidies, essentially a wasteful scheme to transfer money from consumers to corn growers and refiners. Even environmentalists admit that ethanol does not lead to CO_2 reductions overall and has many other undesirable environmental consequences. Among the worst of the consequences of this "biofuel craze" has been the rise in the world price of corn, doubling to $7.50 a bushel in a six-month period in 2010 (EPRF, Inc. 2011). It led to food riots in many developing nations and served to perpetuate poverty throughout the world.

* * *

All in all, the Kyoto Protocol caused nothing but disasters. The billions and even trillions of dollars in financial subsidies it unleashed established politically important stakeholders who continue to fight for "renewable energy," "sustainable development," and other such programs, all in the name of "saving the Earth's climate for our children and grandchildren."

3

The Paris Agreement

THE 1997 KYOTO Protocol expired in 2015, after surviving fifteen years, mostly spent on life support. It reached its peak at COP-13 held in Bali in 2007, had a sudden unexpected collapse in Copenhagen in 2009, and was in a coma after that. In 2015, at COP-21 in Paris, a new agreement called the Paris Agreement was reached, but it was not a treaty (at least not in the United States) and at the time of this writing (in late 2019) it too is on life support.

Failure in Copenhagen

The end of the Kyoto misadventure became evident in 2009 at COP-15 in Copenhagen. Even desperate efforts by scientist-alarmists failed to make an impact. Last-minute fake science reports from environmental interest groups, such as the *Copenhagen Diagnosis* (Allison et al. 2009), could not overcome political resistance. China and major developing nations rejected all efforts to impose limits on their use of fossil fuels; economic growth today proved to be more important than hypothetical climate disasters decades or centuries in the future.

European leaders in Copenhagen hoped to isolate China and India, the main opponents of an agreement, by insisting that they commit to reduce their emissions as a condition for an agreement. However, US President Barack Obama, eager to return to the United States to announce passage of his health care initiative, broke ranks with other developed countries and prepared to leave the conference without an agreement. At the last minute, he negotiated a face-saving but toothless agreement with the leaders of the BASIC (Brazil, South Africa, India, and China) group of countries that became the Copenhagen Accord. The story is told well by Rupert Darwall in his book, *The Age of Global Warming: A History*.

A leak of emails from the University of East Anglia's Climate Research Unit (CRU), an incident labeled "Climategate" described in greater detail in Chapter 6, may also have played a decisive role in the collapse of negotiations in Copenhagen. Not only did a clique of key IPCC scientists hide their raw temperature data and the methodology of their selection and adjustments, but they conspired to delete incriminating emails and fought hard against all attempts by independent outside scientists to replicate their results. They also undermined the peer-review system and tried to make it impossible for skeptical scientists to publish their work in scientific journals.

The Paris Agreement

The parties to the FCCC convened yet again, for the twenty-first time, in Paris in 2015 and finally negotiated a treaty that would replace the Kyoto Protocol. Called COP-21, the meeting attracted the usual cast of characters, delegates from nearly 200 nations who had made a lifetime career out of the climate business plus some 15,000 hangers-on. I predicted at the time (see Singer 2015) that they would fail to reach an *effective* international agreement for a variety of reasons: important developing countries have other priorities; scandals are brewing and may flare up; and the climate itself is not cooperating. But I also predicted the media would portray Paris as a huge success, trying to burnish the environmental-climate legacy of President Barack Obama. I was right on both counts.

By midcentury, GHG emissions by the United States are likely to be less than 10 percent of the world total and thus of little consequence. Yet Obama committed the United States to strict emission reductions by 2030: 32 percent with respect to 2005 emission levels. Obama also pledged $3 billion from US taxpayers to the United Nations' Green Climate Fund by 2020 and endorsed an even more ambitious goal of having developed countries eventually pay $100 billion a year to help developing countries mitigate climate change. When challenged by Russia on his leadership in the Middle East, Obama replied (on *60 Minutes*, on October 11, 2015): "My definition of leadership would be leading on climate change, an international accord that potentially we'll get in Paris."

Obama actively pushed other nations to make commitments to cut GHG emissions, and most obliged him by making meaningless commitments that will have very little effect on actual levels of GHGs and even less on the world climate. China, for example, agreed to peak its emissions in 2030, but it would do nothing to stem growth in the fifteen years leading up to that deadline. Chinese officials calculated, apparently, that by then their population and demand for electric power will have stabilized. In other words, their "commitment" involved no real hardships.

Similarly, in a half-hearted commitment, India pledged to peak its emissions sometime around the middle of the century. However, India's actual plan, announced in 2019, is to double its domestic coal production over the next five years and then continue to use fossil fuels to generate the electricity that is badly needed by its population (*Business Today* [India] 2019). Meanwhile, Norway is massively expanding its oil production (Holter 2019). In Europe, eastern nations will continue to build coal-fired power plants (Ekblom 2019). Even Germany is turning to coal, having foolishly decided, after the Fukushima nuclear disaster, to phase out its well-operating nuclear reactors.

Just as Climategate may have helped derail negotiations in Copenhagen six years earlier, the days leading up to COP-21 saw a scandal that diminished public support for an agreement in the United States. George Mason University professor Jagadish Shukla, founder of the Institute of Global Environment and Society, was accused of receiving $63 million in US government funds, much of it flowing into his and family members' pockets (Mooney 2016). His downfall came when he organized a very public campaign against scientific skeptics, accusing them and their financial supporters of bad faith and profiteering. Some thought to ask the question: Where does Prof. Shukla get *his* funding? The dirty laundry was on full display in congressional hearings organized by Rep. Lamar Smith (R-TX), chairman of the House Committee on Science.

This scandal may have helped convince the public that worries about climate change are driven mostly by money. Other examples come to mind: the promised $100 billion/year subsidy (bribe?) to developing nations (including

China!); Solyndra and a plethora of other "clean" energy projects that went bankrupt, costing taxpayers billions of dollars in wasted subsidies; Al Gore's rise to become a centimillionaire by buying and selling fake "carbon credits"; and many more. Why have we spent some $25 billon on climate science just in the past decade if the "science is settled"?

A Big "Nothing Burger"

COP-21 did produce an agreement—the Paris Agreement, officially the successor to the Kyoto Protocol—but not an effective one. The accord can be briefly summarized as follows:

- Each nation proposes to reduce emissions but sets its own *voluntary* emission target for GHGs, especially for CO_2; there is no overall target for global reduction.
- Each nation reports its own emissions; there is no overall supervision.
- No sanctions are applied if a country fails to abide by its announced plan.

The agreement followed the pattern of the US–China agreement of November 2014, in which China decided to continue with business as usual until reaching a 2030 peak, and only then gradually reducing its emissions. In this manner, each signatory nation can pick and choose its own emission target and time line.

Unlike the Kyoto Protocol, the Paris Agreement has little to do with climate. The accord is mainly about money transfers and virtue signaling, designed to provide a legacy for Obama. Although it talks bravely about keeping global warming below 2°C, it never explains how to define and measure this (alleged) "critical" threshold. It is simply a scheme for redistributing money from Western nations to developing countries, funding the IPCC and other UN bureaucracies, and *possibly* reducing emissions of GHGs. The current plan is to revisit and attempt to tighten national commitments every five years.

Obama agreed to and even sought this language because he knew the US Senate would not approve a binding climate treaty. Even after twenty years, everyone in Washington, DC, still remembers the unanimous Senate vote for the Byrd-Hagel Resolution (of July 1997) against such a treaty. The Obama administration planned to meet US commitments though executive orders and anti-fossil fuel regulations promulgated by the US Environmental Protection Agency (EPA). But on February 8, 2016, the US Supreme Court effectively killed the EPA's Clean Power Plan (CPP), the centerpiece of the US commitment to meet the goals of the Paris Agreement. A year and a half after that, on June 1, 2017, President Donald Trump announced that the United States would cease all participation in the accord. On November 4, 2019 (the earliest date allowed by signatories to the accord), Trump followed through on his promise by notifying the FCCC that the United States would formally withdraw from the agreement.

Trump was right to withdraw the United States from the Paris Agreement. Like the Kyoto Protocol, it lacked any scientific justification, would have wasted billions of dollars, and was unfairly biased against American interests. I predicted COP-21 would fail to produce an effective climate treaty, and I was right on all counts. It was, as they say, a big "nothing burger." Of course, environmental groups and their stenographers in the media roundly condemned the president for "ignoring climate science" and "recklessly endangering the world." But on this issue, Trump understood the science, economics, and politics better than any of his critics.

Leaving the FCCC

President Trump should have immediately gone one step further and withdraw the United States from the FCCC. Article 25 says, "At any time after three years from the date on which the Convention has entered into force for a Party, that Party may withdraw from the Convention by giving written notification to the Depositary."

So long as the United States remains a party to the FCCC, it remains committed to the goal, stated in Article 2, "to achieve . . . stabilization of GHG concentrations in the atmosphere at a level that would prevent dangerous anthropogenic interference with the climate system." This opaque and misleading language is the basis for calls for GHG emissions to not just be held level but be cut back, globally, by 60 to 80 percent. This means cutting fossil fuel burning and energy use by corresponding amounts. In the United States, such constraints on energy use would cause severe economic damage and destroy millions of jobs. Internationally, such limits would strangle growth and development for most of the world's population and make them forever dependent on foreign aid. Even now, the FCCC mandates money transfers to developing nations, thereby removing such decisions from the normal legislative process.

Despite the withdrawal of the United States, other nations have vowed to implement the Paris Agreement and some US states, municipalities, and even universities have pledged to force their residents to reduce their emissions by 32 percent to keep Obama's old promise. There will be claims forthcoming that the Paris Agreement is *already effective* in slowing down global warming, reducing extreme weather events, etc. Don't believe any such claims.

* * *

As this brief history reveals, concern over the possible impact of human activity on the Earth's climate is not new. Theories of how CO_2 emissions caused by the combustion of fossil fuels might cause warming or cooling date back to the early nineteenth century but were generally not taken seriously until a century later, in the late 1980s and early 1990s. Contrary to some accounts of the history of the scientific debate, there was no gradually emerging "consensus" on the human role in climate change. Rather, politics quickly overtook science as environmental advocates and other interest groups recognized the utility of the climate change issue in advancing their own agendas.

Important questions concerning attribution and our inability to observe and measure climate processes were raised by scientists early on and were left unresolved while voices in politics and the media stampeded a worried public into embracing a radical and unnecessary plethora of taxes, regulations, and subsidies in the name of "stopping global warming." So much hot talk—and so little cold science.

4

Misled by the IPCC

A **KEY ENABLER** of the push for a global climate treaty, and the source of much of the misinformation and alarmist rhetoric amplified by environmental groups and media around the world, is the United Nations' IPCC. It was created in 1988 by two United Nations agencies, the WMO and the UNEP, the same year James Hansen testified to the US Congress that he was "99 percent" certain that climate change was here.

A Fake Consensus of Scientists

Through a series of well-publicized reports—coauthored by teams of scientists and policy specialists and then edited by government authorities and environmental activists—the IPCC claimed to represent the "consensus of scientists." Actually, it represents the consensus of *politicians* including not only those representing Western democracies but also the world's worst tyrants and dictators. Politicians set the organization's agenda, name the scientists who are allowed to participate, and rewrite the all-important "Summaries for Policymakers" that the vast majority of policy makers and opinion leaders rely on to understand the thick and highly technical full reports. In these ways, the IPCC is a political, not a scientific, organization. This should be apparent from its name, which includes the word "intergovernmental" but not "scientific" or even "international."

The IPCC produced its first scientific assessment (AR1) in 1990; it was updated in 1992 in time for the Rio de Janeiro meeting. (To save space and avoid confusion, the IPCC's five assessment reports are identified in the text and the references as "AR1, "AR2," etc. Each report consisted of contributions by three "working groups," which are abbreviated in the text as WGI, WGII, and WGIII.) AR1's conclusion—that observed temperature changes and the changes calculated by GCMs are "broadly consistent"—is no longer accepted.

The IPCC's second assessment (AR2), released in 1996, arrived at twin conclusions: that the climate changes of the past century are "unlikely to be due entirely to natural fluctuations" and that "the balance of evidence suggests a discernible human influence on global climate." These phrases, while appearing cautious and unobjectionable, misrepresented the findings of the study itself where one reads, "To date, pattern-based studies have not been able to quantify the magnitude of a greenhouse gas or aerosol effect on climate" (IPCC AP2, 434).

Concerning the consistency of GCMs and observations, IPCC AR2 claimed, "The main conclusion that can be drawn from these investigations is that the observed record of global mean temperature changes can be well simulated by a range of combinations of forcing. Best fits are obtained when anthropogenic forcing factors are included, and, when this is done, most of the observed trend is found to result from these factors. Within the range of forcing and model parameter uncertainties, there is no inconsistency between observations and the modelled global mean response to anthropogenic influences" (IPCC AR2, 423). This, too, was a lie.

Corruption of the Peer-Review Process

AR2 marked the beginning of the overt politicization of IPCC reports, a pattern that is most apparent in the Summaries for Policymakers (SPMs), which are altered to meet the needs of the governments that are the members of the IPCC. Scientists who participated in the production of IPCC reports have repeatedly complained about this violation of the scientific method and even resigned in protest (e.g., Landsea 2005; Lindzen 2012; Tol 2014; Stavins 2014).

Commenting on AR2, Frederick Seitz wrote, "But this report is not what it appears to be—it is not the version that was approved by the contributing scientists listed on the title page. In my more than 60 years as a member of the American scientific community, including service as president of both the National Academy of Sciences and the American Physical Society, I have never witnessed a more disturbing corruption of the peer-review process than the events that led to this IPCC report" (Seitz 1996). Box 3 at the end of this chapter presents the full text of Seitz's remarkable description of this scientific fraud.

According to Seitz, "more than 15 sections in Chapter 8 of the report—the key chapter setting out the scientific evidence for and against a human influence over climate—were changed or deleted after the scientists charged with examining this question had accepted the supposedly final text. Few of these changes were merely cosmetic; nearly all worked to remove hints of the skepticism with which many scientists regard claims that human activities are having a major impact on climate in general and on global warming in particular" (Seitz 1996, 16).

One can trace the text changes to a letter of instruction from the US Department of State addressed to Sir John Houghton, head of the IPCC Working Group I, and signed by a Mr. Day Mount, acting deputy assistant secretary of state, environment and development. The final paragraph of the letter reads: "It is essential that the chapters not be finalized prior to the completion of discussions at the IPCC WG I plenary in Madrid, and that chapter authors be prevailed upon to modify their text in an appropriate manner following discussion in Madrid" (Masood 1996). An editorial in the June 13 issue of *Nature* stated that the changes made to Chapter 8 were designed to "ensure that it conformed" to the political agenda of setting up international controls on energy use (*Nature* 1996).

Regrettably, instead of acknowledging the abuses that Seitz and *Nature* reported and then holding accountable the scientists and activists who were responsible, the leaders of the IPCC attacked Seitz and came to the defense of Benjamin Santer, the report's lead author. In September 1996, an "open letter of support" addressed to Santer and signed by some two dozen scientists on behalf of the Executive Committee of the American Meteorological Society and the trustees of University Corporation for Atmospheric Research appeared in the *Bulletin of the American Meteorological Society* (Avery et al. 1996). Together with eleven other scientists, my comments on the open letter appeared in the same journal in January 1997 (Singer et al. 1997). After reciting and documenting the edits and deletions done after peer review and just prior to publication, we warned that "the principal conclusion derived from chapter 8—that 'the balance of evidence suggests a discernible human influence on global climate'—is being misused by politicians." "The real issue then," we concluded, "is the political misuse of the IPCC report and of climate science rather than the ongoing debate about procedure. We urge that this serious matter be energetically addressed by the AMS and by UCAR forthwith."

Alas, the American Meteorological Society (AMS) and University Corporation for Atmospheric Research (UCAR) did not address the corruption that had occurred at the highest levels of the IPCC. As Dennis Avery and I wrote in 2008, "Santer single-handedly reversed the 'climate science' of the whole IPCC report and with it the global warming political process. The 'discernible human influence' supposedly revealed by the IPCC has been cited thousands of times since in media around the world, and has been the 'stopper' in millions of debates among nonscientists" (Singer and Avery 2008, 121).

In 2000, the Hoover Institution published my detailed critique of AR2 and the politics of the Kyoto Protocol as a monograph in a series titled *Essays in Public Policy*. I concluded,

> The science is fairly straightforward. Even if one were to trust the model predictions of future temperature rise, Kyoto is not the way to go: too expensive and quite ineffective. If it is decided that the Climate Treaty (FCCC) calls for limits to CO_2 in the atmosphere, sequestration may be the better alternative for mitigation—at least as an adjunct to the emission controls of the Kyoto Protocol. (Current research suggests that fertilizing the oceans with iron, a micronutrient, may become a cost-effective method.) But the main message from science is that we have already seen high temperatures in the historic climate record; and further, we can be fairly sure that a little warming will restrain sea level rise—not accelerate it and that severe storms and even hurricanes will not increase. Economics also paints a benign picture of global warming. If the latest analyses are borne out, then more warming is what we need—to increase GNP and prosperity (Singer 2000, 39).

The next IPCC Assessment Report, AR3, was released in 2001. This report featured the notorious "hockey stick" graph, invented by Michael Mann, which appeared to erase the Medieval Warm Period (MWP, a period from approximately 950 to 1250) from the historical temperature record by showing little temperature change for a thousand years followed by a sharp rise in the twentieth century. Problems with the hockey stick graph and the fraud that occurred partly to cover up those problems are described later in this chapter and in Chapter 5.

Less widely remarked upon is that the authors of AR3 candidly acknowledged that the limited understanding of climate processes necessarily makes climate modeling an uncertain exercise: "In sum, a strategy must recognize what is possible. In climate research and modeling, we should recognize that we are dealing with a coupled nonlinear chaotic system, and therefore that long-term prediction of future climate states is not possible" (IPCC, AR3, WGI, 774).

In 2002–3, in response to the release of the initial drafts of the IPCC's Fourth Assessment Report (AR4), I convened a group of scientists to independently review the scientific evidence. In 2007, when the final report was issued, we decided to call ourselves the Nongovernmental International Panel on Climate Change (NIPCC) and began publishing our own reports. A list of NIPCC's publications to date appears in Box 4 (Chapter 7).

The IPCC's Fourth Assessment Report (AR4) was released in 2007. It basically doubled down on its reliance on computer models, saying in effect that the climate models that run without a CO_2 effect do not predict the warming shown in the surface temperature record. Therefore, AR4 concludes, CO_2 must be responsible for the difference between known natural forcings and the temperature record. There were two things wrong with this reasoning. First, the surface temperature record is badly flawed and shows a spurious warming trend. Second, this reasoning is circular, since it begins by assuming that all natural forcings are known (with precision) and then concludes that whatever forcings the models do not include must be the result of man-made interference with the climate. But the null hypothesis is that natural forcings account for *all* of the observed warming, and simply *assuming* the null hypothesis is wrong does not prove the greenhouse warming theory is correct.

The Missing Hotspot

Crucially, the IPCC's models fail to produce warming *in the right places*. This failure has come to be known as the "missing hotspot," evidence of which appeared (but was not reported as such) in a US Climate Change Science Program report published in 2006 titled *Temperature Trends in the Lower Atmosphere: Steps for Understanding and Reconciling Differences* (Karl et al. 2006). (See Figure 3.)

All IPCC GCMs show an amplification of trends in the tropical zone, with a "hotspot" in the upper troposphere, while the temperature data from radiosondes (both the Hadley Centre analysis and the Radiosonde Atmospheric Temperature Produces for Assessing Climate [RATPAC] analysis by NOAA) do not show this feature. This "potentially serious inconsistency" was pointed out in a paper I coauthored with David Douglass, John Christy, and Benjamin Pearson in the *International Journal of Climatology* (Douglass et al. 2008). The key graphic in that article, showing the major disparity between models and observations, appears as Figure 4.

Santer and colleagues responded to Douglass and colleagues with an article published in the same issue of *International Journal of Climatology,* offering "new observational estimates of [tropical] surface and tropospheric temperature trends" and concluding that "there is no longer a serious discrepancy between modelled and observed trends" (Santer et al. 2008, 1703–22). But the "new observational estimates" conflict with satellite data. Their modeled trends are an artifact, merely reflecting chaotic and structural model uncertainties that had been overlooked. Thus, the conclusion of "consistency" is not supportable and does not validate model-derived projections of dangerous AGW. For further discussion of this controversy, see Singer (2011, 2013).

Walking Back Key Alarmist Claims

The IPCC's fifth and latest assessment report (AR5) was released in three volumes in 2013 and 2014. The Summary for Policymakers (SPM) of the first volume, the Working Group I report subtitled "The Physical Science Basis," claimed "warming of the climate system is unequivocal, and since the 1950s, many of the observed changes are unprecedented over decades to millennia.

a. Model Predictions

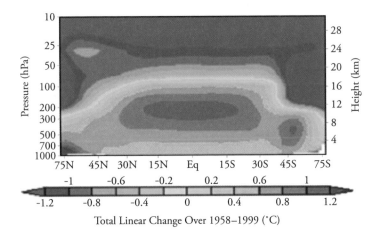

Total Linear Change Over 1958–1999 (°C)

b. Observations

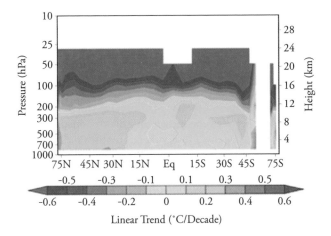

Linear Trend (°C/Decade)

Source: Karl et al. 2006.

Figure 3. The missing hotspot. Figure (a) is Figure 1.3F from Karl et al., 2006, 25, showing GCM predicted temperature trends versus latitude and altitude. Note the increasing trends in tropical midtroposphere, with a maximum around 10km. Figure (b) is Figure 5.7E from the same source, 116, showing observed temperature trends (HadAT2 radiosonde data). Note the absence of increasing trends (i.e., no "hotspot") in tropical midtroposphere.

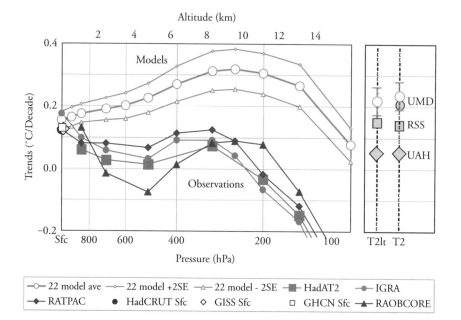

Source: Douglass et al. 2008.

Figure 4. Model and observational temperature trends versus altitude in the tropics. Plot of temperature trend (°C/decade) against pressure (altitude) at thirteen altitude levels between the surface and the tropopause for the period 1979–99. Observations are three independent surface and the average of four radiosonde results, plotted as curves, and five satellite values (two for the lower troposphere [T2lt] and three for midtroposphere [T2]) shown in the right panel. The curve labeled "22 model ave," is the average of the twenty-two climate models run between one and nine simulations to remove the El Niño/Southern Oscillation (ENSO) effect. "22 model +2SE" and "22 model – 2SE" are the uncertainty limits of the mean, meaning one would expect that the mean would lie between these limits with 95 percent probability (i.e., ± two standard deviations). Note that all of the observational values are less than the synthetic values from the models. Only the satellite value from the University of Maryland (UMD) is even within the uncertainty of the model calculation.

The atmosphere and ocean have warmed, the amounts of snow and ice have diminished, sea level has risen, and the concentrations of greenhouse gases have increased" (AR5 2013, 4). Trying to top the ever-rising expressions of confidence in their opinions that appeared in past ARs, the authors wrote, "It is *extremely likely* that human influence has been the dominant cause of the observed warming since the mid-20th century" (AR5 2013, 17).

AR5 raised the bar for errors, exaggerations, and misleading assertions in the climate-change debate. Future chapters will expose and correct many of its false claims, but here I will focus on something in AR5 that is widely overlooked by politicians and the news media: AR5 actually *walked back* some key alarmist claims promulgated in previous IPCC reports or by scientists prominently associated with the IPCC. The following quotations in italics are from the IPCC's 2013 SPM, and the text following them is based on a paper I wrote with several coauthors published by the NIPCC in 2013 (Idso et al. 2013).

1. *"The rate of warming over the past 15 years (1998–2012; 0.05 °C per decade), which begins with a strong El Niño, is smaller than the rate calculated since 1951 (1951–2012; 0.12 °C per decade)" (p. 5).*

The IPCC conceded for the first time that a 15-year period of no significant warming occurred since 1998 despite a 7 percent rise in atmospheric CO_2 levels. It also acknowledged that on a longer time scale the rate of global warming has decelerated since 1951, despite an accompanying 80 ppm or 26 percent increase in CO_2 levels (from 312 ppm to 392 ppm). (Since AR5 was published in 2013, atmospheric CO_2 has risen to about 412 ppm in 2019, about 130 ppm or 46 percent higher than the preindustrial level of 280 ppm.)

2. *"Continental-scale surface temperature reconstructions show, with high confidence, multi-decadal periods during the Medieval Climate Anomaly (year 950–1250) that were in some regions as warm as in the late twentieth century" (p. 5).*

AR3 and AR4 previously argued that the magnitude of the late twentieth-century global warming exceeded that of the MWP. The IPCC now abandoned and repudiated this claim and along with it the notorious "hockey stick" featured in AR3 and still visible in AR4. Importantly, this means the IPCC cannot rule out natural variability as the cause of the twentieth century's warming.

3. "It is very likely that the annual mean Antarctic sea ice extent increased at a rate of 1.2 to 1.8 percent per decade (range of 0.13 to 1.07 million km² per decade) between 1979 and 2012" (p. 9).

GCMs predict that GHG forcing would cause surface warming and ice melting simultaneously in both north and south polar regions. While the Arctic has seen rising temperatures and declines in ice sheet mass and sea ice, Antarctica has seen falling temperatures, its ice sheet (taken as a whole) has probably been stable or even increased in mass in recent years, and sea ice in that part of the world has been *increasing* (Fountain et al. 2017; Engel et al. 2018). There is no *a priori* reason why increasing atmospheric CO2 would cause the Arctic to warm but the Antarctic to cool, or for Arctic ice to melt but Antarctic ice to remain stable or even increase. It is a welcome advance that the IPCC acknowledged the facts relevant to this matter.

4. "There are, however, differences between simulated and observed [global temperature] trends over periods as short as 10 to 15 years (e.g., 1998 to 2012)" (p. 15).

Earlier drafts of the SPM (before it was edited by politicians and activists) contained stronger language regarding the failure of GCMs, saying "*Models do not generally reproduce the observed reduction in surface warming trend over the last 10–15 years.*" Either way, the IPCC admitted that its climate models failed to predict the lack of warming over the past fifteen years (1998–2013). Elsewhere in the SPM, the IPCC admits to "*low confidence in the representation and quantification of [cloud and aerosol] processes in models*" (p. 16) and "*most models simulate a small downward trend in Antarctic sea ice extent, albeit with larger inter-model spread, in contrast to the small upward trend in observations"* (p. 16). These statements represent a significant reduction of confidence in the IPCC's models, which are at the heart of its claim to be able to predict future climate conditions.

5. "Equilibrium climate sensitivity is likely in the range 1.5°C to 4.5°C . . . " (p. 16) and "No best estimate for equilibrium climate sensitivity can now be given because of a lack of agreement on values across assessed lines of evidence and studies" (p. 16, n. 16).

Equilibrium climate sensitivity (ECS) is the amount of warming expected to result from a doubling of atmospheric CO_2 as the climate system tends toward equilibrium (>1,000 years). The IPCC's AR4 stated a range of 2.0°C to 4.5°C for ECS. By reducing the ECS lower limit to 1.5°C, the IPCC in AR5 conceded that less certainty exists than in 2007. Indeed, the climate sensitivity of atmospheric CO_2 is now as uncertain as it was in 1979 when the Charney report established the same range. In other words, no refinement has been made in thirty-four years in determining how much warming is likely to result from a doubling of atmospheric CO_2.

The decision not to designate a "best estimate" for ECS is unique in IPCC's history and a further indication of growing uncertainty. It probably reflects the publication of a number of then-recent papers (e.g., Aldrin 2012; Ring et al. 2012; Lewis 2013) in which sensitivity was estimated from observations to be between 1.2°C and 2.0°C, a range that extends below the IPCC's estimates.

6. "The transient climate response is likely in the range of 1.0°C to 2.5°C (high confidence) and extremely unlikely greater than 3°C" (p. 16).

Transient climate response (TCR) is the amount of warming expected to result from a doubling of atmospheric CO_2 after seventy years, given a rate of CO_2 increase of 1 percent per year. By reducing the bottom of the range of TCR to 1.0°C, the IPCC's estimate of human-caused warming for the rest of the twenty-first century now overlaps with those of many independent scientists who put the response in the range of 0.3°C to 1.2°C. In setting the top of the range at 3.0°C, the IPCC's estimate now falls within the range of natural climate variation over the past 6 million years. Because it falls within the warm natural temperature limit that planet Earth has attained recently, any such change (should it actually happen) is unlikely to be "dangerous."

7. *"It is very unlikely the AMOC [Atlantic Meridional*
   Overturning Circulation] will undergo an abrupt
   transition or collapse in the 21st century for the scenarios
   considered" (p. 24).

The IPCC also had indicated in its AR4 report that it was unlikely the AMOC would collapse due to fresh water input to the ocean from melting ice. However, this did not prevent IPCC- related scientists and environmental lobbyists from arguing in the interim that increasing GHGs might cause such a calamity. There is no evidence linking rising CO_2 to abrupt climate change, whether via this avenue or any other. The IPCC now recognized this.

8. *"Global mean sea level rise for 2081–2100 will likely be*
   in the ranges of 0.26 to 0.55 m for RCP2.6, 0.32 to 0.63 m
   for RCP4.5, 0.33 to 0.63 m for RCP6.0 and 0.45–0.82 m
   for RCP8.5 (medium confidence)" (p. 25).

RCP2.6, RCP4.5, RCP6.0, and RCP8.5 are four "representative concentration pathways" named after radiative forcing (RF) values of cumulative anthropogenic CO_2 and CO_2 equivalents by the year 2100 relative to pre-industrial values (2.6, 4.5, 6.0, and 8.5 W/m^2, respectively). While the IPCC does not attach probabilities to any of these scenarios, RCP4.5 seems most credible. Even the IPCC's lowest estimate of a 26 cm (10") rise by 2100, for RCP2.6, is significantly above the 18 cm (7") rise suggested by many independent scientists (Parker and Ollier 2017). The highest estimate of 82 cm (32") by 2100 falls well below the 1.4 m (55") promulgated by IPCC-related scientists such as Rahmstorf (2007) and others. Overall, these sea level projections are still high when compared to observed trends and the best estimates reported in the scientific literature.

9. *"Low confidence" in "increases in intensity and/or duration of drought" and "increases in intense or tropical cyclone activity" (p. 7, Table SPM.1).*

Many papers by IPCC-related scientists, and also previous Assessment Reports, argued that CO_2 forcing would result in dangerous increases in the magnitude or frequency of extreme climatic events including cyclones and droughts. By admitting it has "low confidence" in predictions of more frequent or more extreme droughts and tropical cyclones, IPCC is specifically revoking its previous more alarmist claims. This is for good reason: research shows no increases in drought (Kleppe et al. 2011) or tropical cyclone activity (Klotzbach et al. 2018) during the past forty years.

A Huge Public Deception

These nine walk-backs in the IPCC's latest report were not featured in news releases or underscored in the SPM. Just the opposite: they were largely buried under new rhetoric and false claims. They suggest that the IPCC was not entirely honest in its past reports and is only grudgingly admitting its mistakes today . . . while making new ones. But this hardly scratches the surface of the huge public deception that occurred in the three decades since the first IPCC report was released. Chapter 5 describes the "hockey stick" scandal, in which a small group of scientists conspired to rewrite climate history in order to claim the temperature increases in the twentieth century were "unprecedented," and Chapter 6 describes "Climategate," an episode in which prominent IPCC-affiliated scientists were found to be conspiring to hide data from other scientists and prevent the publication of competing views in scientific journals. The IPCC's other major errors and oversights will be reported in Part 2 of this book.

The ongoing negotiations by the UN Conferences of Parties take for granted that the science of climate change is "settled." The media and many politicians and environmental activists parrot this claim. *Nothing could be further from the truth.* The climate models that predict a major warming in the next century have not been validated by observations and therefore cannot—and should not—be used as a basis for decision-making. In spite of the constant use of the phrase "scientific consensus," there is substantial disagreement on many issues within the community of atmospheric scientists and climate specialists. Indeed, most scientists believe that the climate change issue should be considered "unfinished business" requiring much further research.

* * *

In conclusion, the IPCC misled an entire generation of scientists and policymakers, telling them the human impact on the Earth's climate poses a genuine threat to human well-being and other life on the planet while deliberately and repeatedly hiding uncertainty, the absence of critical data, and evidence that questions or contradicts its apocalyptic prediction. Many thoughtful and well-intended people accept the IPCC's claims unconditionally, taking at face value its claim to represent the "consensus of scientists." They were betrayed. The result is a terrible crime against science, the adoption of unnecessary and very costly public policies, and grave damage to the reputation and credibility of science.

Box 3

"A Major Deception on Global Warming"
By Frederick Seitz
Wall Street Journal
June 12, 1996

Last week the Intergovernmental Panel on Climate Change [IPCC], a United Nations organization regarded by many as the best source of scientific information about the human impact on the Earth's climate, released "The Science of Climate Change 1995," its first new report in five years. The report will surely be hailed as the latest and most authoritative statement on global warming. Policy makers and the press around the world will likely view the report as the basis for critical decisions on energy policy that would have an enormous impact on U.S. oil and gas prices and on the international economy.

This IPCC report, like all others, is held in such high regard largely because it has been peer-reviewed. That is, it has been read, discussed, modified and approved by an international body of experts. These scientists have laid their reputations on the line. But this report is not what it appears to be—it is not the version that was approved by the contributing scientists listed on the title page. In my more than 60 years as a member of the American scientific community, including service as president of both the National Academy of Sciences and the American Physical Society, I have never witnessed a more disturbing corruption of the peer-review process than the events that led to this IPCC report.

A comparison between the report approved by the contributing scientists and the published version reveals that key changes were made after the scientists had met and accepted what they thought was the final peer-reviewed version. The scientists were assuming that the IPCC would obey the IPCC Rules—a body of regulations that is supposed to govern the panel's actions. Nothing in the IPCC Rules permits anyone to change a scientific report after it has been accepted by the panel of scientific contributors and the full IPCC.

The participating scientists accepted "The Science of Climate Change" in Madrid last November; the full IPCC accepted it the following month in Rome. But more than 15 sections in Chapter 8 of the report—the key chapter setting out the scientific evidence for and against a human influence over climate—were changed or deleted after the scientists charged with examining this question had accepted the supposedly final text.

Few of these changes were merely cosmetic; nearly all worked to remove hints of the skepticism with which many scientists regard claims that human activities are having a major impact on climate in general and on global warming in particular.

The following passages are examples of those included in the approved report but deleted from the supposedly peer-reviewed published version:

- "None of the studies cited above has shown clear evidence that we can attribute the observed [climate] changes to the specific cause of increases in greenhouse gases."

- "No study to date has positively attributed all or part [of the climate change observed to date] to anthropogenic [man-made] causes."

- "Any claims of positive detection of significant climate change are likely to remain controversial until uncertainties in the total natural variability of the climate system are reduced."

The reviewing scientists used this original language to keep themselves and the IPCC honest. I am in no position to know who made the major changes in Chapter 8; but the report's lead author, Benjamin D. Santer, must presumably take the major responsibility.

IPCC reports are often called the "consensus" view. If they lead to carbon taxes and restraints on economic growth, they will have a major and almost certainly destructive impact on the economies of the world. Whatever the intent was of those who made these significant changes, their effect is to deceive policy makers and the public into believing that the scientific evidence shows human activities are causing global warming.

If the IPCC is incapable of following its most basic procedures, it would be best to abandon the entire IPCC process, or at least that part that is concerned with the scientific evidence on climate change, and look for more reliable sources of advice to governments on this important question.

5

The Hockey Stick Deception

THE "HOCKEY STICK" graph was invented by Michael Mann, who received his PhD in geology and geophysics in 1998, and in 1999 was named a research assistant professor at the University of Massachusetts. His graph first appeared in articles published in 1998 and 1999 coauthored with two more senior colleagues (Mann, Bradley, and Hughes 1998; 1999) and then was featured in the Third Assessment Report of the IPCC published in 2001. (See Figure 5.)

Mann's graph enabled the IPCC and the Clinton administration to respond to the argument that natural climate variations exceed whatever effect human activity might have had in the twentieth century by claiming even the very biggest past historic changes in temperatures *simply never happened*. Despite Mann's junior status in the scientific community, the IPCC prominently displayed the hockey stick in its Third Assessment Report. The Clinton administration featured the graph as the first visual in *Climate Change Impacts on the United States: The Potential Consequences of Climate Variability and Change* (National Assessment Synthesis Team 2001). Mann was named an IPCC lead author and an editor of the *Journal of Climate*, a major professional journal. His graph subsequently appeared in Al Gore's movie, *An Inconvenient Truth*.

The Hockey Stick Illusion

Mann and his colleagues used several temperature proxies (primarily tree rings, in particular one data series created by Keith Briffa of the CRU) as a basis for assessing past temperature changes from 1000 to 1980. They then

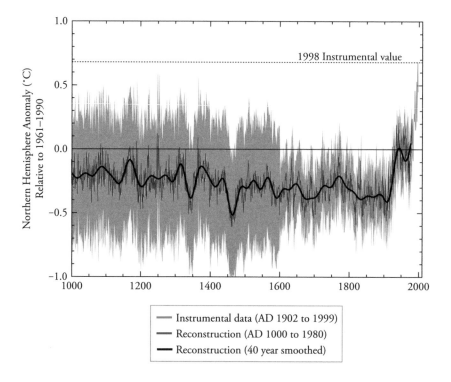

Source: IPCC, AR3, Technical Summary of the Working Group I Report, 29.

Figure 5. The "hockey stick" graph invented by Michael Mann and coauthors as it appeared in IPCC AR3. Millennial Northern Hemisphere (NH) temperature reconstruction (blue—tree rings, corals, ice cores, and historical records) from AD 1000 to 1999 and instrumental data (red) in the twentieth century. Smoother version of NH series (black), and two standard error limits (gray shaded) are shown.

grafted the surface instrument temperature record of the twentieth century onto the pre-1980 proxy record. The effect was visually dramatic. Gone were the difficult-to-explain MWP (around 1100 AD) and Little Ice Age (LIA, about 1400 to 1850), and in their place Mann put 900 years of stable global temperatures—until about 1910. Then the twentieth century's temperatures seem to rocket upward out of control.

The implication of Figure 5, that the warming of the twentieth century was unprecedented or unnatural, is obvious but *wrong*. Instead of using Briffa's proxy record from 1978 to 1998, Mann used instrumental data from surface thermometers, which recorded the super–El Niño warming of 1998. The switch created the appearance of a sharp rise in surface temperatures, but it was an illusion (some would use the word "deception"). (See Christy 2011.) He hid from other researchers his data for the crucial 1978–97 interval and has never revealed them. Doing so would not support his narrative of an "unprecedented" rise in temperature.

The Illusion Exposed

Starting in 2003, Canadian statistician Stephen McIntyre and economist Ross McKitrick began exposing Mann's shoddy analysis. First, they requested the original study data from Mann. It was provided—haltingly and incomplete— indicating that no one else had previously requested the data, revealing the absence of true peer review. They found the data did not produce the claimed results "due to collation errors, unjustifiable truncation or extrapolation of source data, obsolete data, geographical location errors, incorrect calculation of principal components and other quality control defects" (McIntyre and McKitrick 2003, 751).

Using corrected and updated source data, McIntyre and McKitrick recalculated the Northern Hemisphere temperature values for the period 1400–1980 using Mann's own methodology. Their work was published in *Energy & Environment*, with the data refereed by the World Data Center for Paleoclimatology. "The major finding is that the values in the early 15th century exceed any values in the 20th century," they reported (McIntyre and McKitrick 2003, 751). In other words, the Mann study was fundamentally wrong.

Next, they examined Mann and colleagues' statistical methodology, whereby each series of tree ring data was transformed by subtracting mean temperatures during the twentieth century, dividing by the standard deviation of that same series, then dividing the tree ring series again by the standard deviation of the residuals of a linear trend superimposed on the series of mean temperatures during the twentieth century. Use of this algorithm was not reported in Mann's published work. McKitrick and McIntyre demonstrated that even when data without trends are entered into the formula, hockey-stick-shaped patterns result.

McKitrick and McIntyre also discovered that Mann's analysis relied heavily on tree rings taken from ancient bristlecone pine trees from the western United States, even though the growth rate of those trees is known to be determined more by ambient levels of CO_2, a natural fertilizer, than by temperatures. Mann must have realized this, since the source he cited was an academic article specifically making this very point (Graybill and Idso 1993), yet he chose this dataset presumably because it supported his narrative. McIntyre and McKitrick demonstrated that removing the bristlecone pine tree data eliminates the distinctive rise at the end of the "hockey stick."

Willie Soon, David Legates, and Sallie Baliunas (Soon et al. 2004) demonstrated that Mann redrafted the "hockey stick" several times, with each rendition pushing the 2000 value higher to accommodate the assertion that "the 1990s were the warmest decade of the past millennia with 1999 being the warmest year."

Mann and his team were forced to publish a correction in *Nature* admitting to errors in their published proxy data, but they claimed that "none of these errors affect our previously published results" (Mann, Bradley, and Hughes 2004, 105). That claim, too, was contradicted by McIntyre and McKitrick (2005), by statistics expert Edward Wegman (Wegman, Scott, and Said 2006), and by a National Academy of Sciences report (NAS 2006). The NAS skipped lightly over the errors of the hockey-stick analysis and concluded it showed only that the twentieth century was the warmest in 400 years, not 2,000 years as Mann and colleagues had claimed, but this conclusion is hardly surprising, since the LIA was at its coldest 400 years ago. It was the claim that temperatures in the second half of the twentieth century were the highest in the last millennium that properly generated the most attention, and which McIntyre and McKitrick had shown to be unproven.

One can disprove Mann's claim by demonstrating that about 1,000 years ago, there was a world-wide MWP when global temperatures were equally as high as or higher than they were over the latter part of the twentieth century, despite there being approximately 25 percent less CO_2 in the atmosphere than

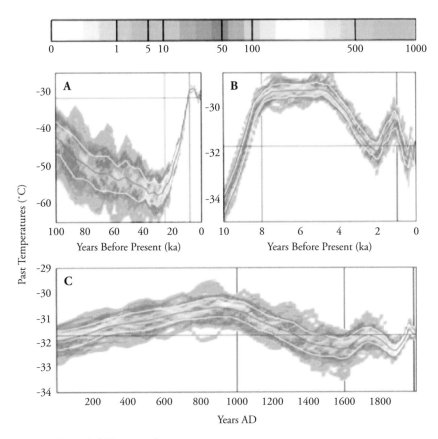

Source: Dahl-Jensen et al. 1998.

Figure 6. Temperature data from the Greenland Ice Core Project (GRIP). (A) Data from present to 100,000 years ago, (B) from present to 10,000 years ago, and (C) from present to 2,000 years ago. Note the pronounced Medieval Warm Period and Little Ice Age. The authors state explicitly: no warming is seen after 1940.

there is today. One example of proxy data that show this, and not Mann's sharp rise in the twentieth century, is shown in Figure 6. Willie Soon and colleagues also demonstrated that "many records reveal that the 20th century is likely not the warmest nor a uniquely extreme climatic period of the last millennium" (Soon et al. 2003, 233). This real-world fact conclusively demonstrates there is nothing unnatural about the planet's current temperature, and whatever warming occurred during the twentieth century could have been caused by the recurrence of whatever cyclical phenomena created the equal or even greater warmth of the MWP.

The degree of warming and climatic influence during the MWP varied from region to region and its consequences were manifested in several ways. But that it occurred and was a global phenomenon is certain; The Center for the Study of Carbon Dioxide and Global Change has analyzed more than 200 peer-reviewed research papers produced by more than 660 individual scientists working in 385 institutions from 40 countries that bear witness to this truth. Many of these studies are described and cited in the *Climate Change Reconsidered* series published by the NIPCC.

* * *

In short, the "hockey stick" is a *fictitious construct* designed to deceive gullible viewers. Its prominence in the climate change debate even today, some twenty years after it entered the literature and was thoroughly debunked, is testimony to the influence of politics on science and the failure of the science community to police itself.

6

The Climategate Scandal

IN NOVEMBER 2009, a whistleblower leaked thousands of emails from the CRU. These emails involved mainly Michael Mann and several other scientists with high positions in the IPCC and exposed their completely unethical attempts to suppress contrary opinions and publications from climate skeptics—unfortunately with considerable success. A second batch of emails was released two years later. More complete discussions of the Climategate emails can be found in *The Hockey Stick Illusion: Climategate and the Corruption of Science* by Andrew W. Montford (Montford 2010) and *The Climategate Emails*, edited and annotated by John Costella (Costella 2010). The full collection of Climategate I and II emails and files can be found online at https://sealevel.info/FOIA/. A more technical discussion has been ongoing in Stephen McIntyre's blog https://climateaudit.org.

"An Organized Conspiracy"

The incident, called "Climategate," exposed the manhandling of fundamental data, such as Mann's data manipulation to "hide the decline" in air temperature post-1980 that would have been revealed by his own proxy data if the instrumental record had not been affixed to the end of the "hockey stick." Commenting on the affair, Larry Bell wrote for Forbes.com, "Many [of the leaked emails] clearly confirm that top IPCC scientists consciously misrepresented and actively withheld important information . . . then attempted to prevent discovery. Included are CRU's Director of Research, Phil Jones; the US National Center for Atmospheric Research (NCAR) climate's analysis section head, Kevin Trenberth; and beleaguered Penn State University 'hockey stick' originator, Michael Mann" (Bell 2011).

Myron Ebell, director of the Competitive Enterprise Institute's Center on Energy and Environment, said at the time, "If there were any doubts remaining after reading the first Climategate e-mails, the new batch of e-mails that appeared on the web today [November 22, 2011] make it clear that the UN Intergovernmental Panel on Climate Change is an organized conspiracy dedicated to tricking the world into believing that global warming is a crisis that requires a drastic response." He went on to say, "Several of the new e-mails show that the scientists involved in doctoring the IPCC reports are very aware that the energy-rationing policies that their junk science is meant to support would cost trillions of dollars" (quoted in Bell 2011).

I also commented on the affair at the time for the Institute of Economic Affairs, a British think tank. I said,

> The Climategate disclosures over the past few days, consisting of some thousand emails between a small group of British and US climate scientists, suggest that global warming may be man-made after all—created by a small group of zealous scientists!
>
> It would seem they have used flawed data, phony statistics, and various "tricks." They appear to have covered up contrary evidence and refused to open their work to the scrutiny of independent scholars. It has also been suggested that by keeping out "intruders," by reviewing their own papers, by capturing scientific journals and intimidating editors, they have tried to suppress dissent.
>
> I do not wish to discuss any of the ethical or legal aspects, which may be self-evident.
>
> I consider the whole matter a great tragedy not only for science but also for the institutions involved and for many of the scientists involved who have in fact spent many years and whole careers on their work. In particular, I have some personal sympathy for Philip Jones and feel he has been dealt a bad hand. Trying to correct temperature observations from weather stations around the world is extremely difficult work. It involves much detail; it is certainly not traditional science. However,

I cannot endorse the actions of this group and hope that an impartial investigation will bring closure to this difficult matter (Singer 2009).

I ended by saying, "Inevitably, the public's view of science will be affected and this will hurt all of science." These were probably the truest words spoken about the scandal. Today, surveys show a considerable decline in public confidence in the proclamations of climate scientists. This is one reason often given when pollsters ask why a person does not believe global warming poses much of a threat.

The Scientific Background

As explained in the previous chapter, Mann's claim to fame derives from his contentious (and now thoroughly discredited) "hockey stick" research papers originally published in *Nature* (1998) and *Geophysical Research Letters* (1999). His idiosyncratic analysis of proxy (nonthermometer) data from sources like tree rings, ice cores, and ocean sediments did away with the MWP and LIA, amply documented by Prof. H. H. Lamb, the founding director of the CRU.

There are many ways to present and interpret the temperature record, so Mann's creation might just be put down as one of many that reflect the subjective judgments of their creators. But Mann's creation was purported to be something more, and so appeared to be a deliberate effort to mislead. It is surprising that the work of such a junior academic—he joined the University of Virginia faculty as an assistant professor in 1999, the year of his first publications—would be featured in the IPCC reports. Mann often misleads readers by mixing up temperature level (i.e., absolute temperature) with temperature trends. While current levels are high (since the climate is still recovering from the LIA), the trend has been essentially zero for more than two decades—in spite of rapidly rising CO_2 concentrations.

In reading Mann's original papers, I noticed his temperature record based on proxy data suddenly stopped in 1978 and was joined smoothly to the thermometer record from surface weather stations, which showed a steep rise in temperature. By contrast, atmospheric temperatures measured from weather satellites show only insignificant warming between 1978 and 1997—as do the independent data from weather balloons around the world. Puzzled by this disparity, I emailed Mann (then at the University of Virginia) and politely asked about his post-1978 proxy temperatures. I got a nasty reply in return, confirming my suspicion that Mann was hiding the data because they disagreed with the widely accepted thermometer record. I believe this is the true meaning of the phrase "Mike's *Nature* trick," used in the leaked Climategate emails—in conjunction with "hide the decline." It all suggests manipulation of crucial data.

International Journal of Climatology

I have a personal connection to the Climategate scandal. A paper I coauthored in 2007 (henceforth "Douglass et al."), the findings of which are reported in Chapter 4, was the subject of some of the emails shared by the conspirators. The episode is important because it clearly exposes a group of climate scientists, many of them closely associated with the IPCC, conspiring to delay the publication of a paper that challenges their favored narrative. My coauthors were David H. Douglass, professor of physics at University of Rochester; John R. Christy, distinguished professor of atmospheric science at the University of Alabama in Huntsville; and Benjamin Pearson, professor of physics at the University of Rochester. Douglass and Christy wrote a detailed account of the email exchanges for *American Thinker*, a website, in 2009. Here is a summary of their account.

Douglass et al. was submitted to the *International Journal of Climatology* (*IJC*) on May 31, 2007 and published online six months later, on December 5, and finally appeared in print on November 15, 2008, eleven months after it appeared online. A paper by Santer et al. attempting to rebut Douglass et al. was submitted on March 25, 2008, seven months *before* the print version of our article appeared, was published online on October 10, one month *before* our article appeared in print, and appeared in print on November 15, in the *same issue* of *IJC* as our article and only thirty-six days after online publication. The leaked emails reveal how the normal conventions of the peer-review process were compromised by a team of global warming scientists whose members included Michael Mann, Ben Santer, Phil Jones, Timothy

Osborn, Tom Wigley, and the senior editor of the *IJC*, Glenn McGregor. The conspirators delayed publication of our paper by nearly a year to prepare and publish a paper rebutting it.

The leaked emails begin with Andrew Revkin, a reporter for the *New York Times*, sending three team members an email (November 30, 2007) with the page proofs of the Douglass et al. paper. This is a week *before* the online publication. Revkin was sharing a leaked version of the paper to scientists, expecting them to help him write a critical review of the article, revealing that this was a well-established pattern for dealing with new research questioning the AGW theory.

A series of emails followed aimed at allowing Santer et al. to get a rebuttal published without giving Douglass et al. "the opportunity to have a response." Tim Osborn, a colleague of Jones at the CRU and a member of the editorial board of *IJC*, writes (January 10, 2008) that he has contacted the editor, Glenn McGregor, to "see what he can do." According to Osborn, McGregor "promises to do everything he can to achieve a quick turn-around." He also says "[McGregor] may be able to hold back the hardcopy (i.e., the print/paper version) appearance of Douglass et al., possibly so that any accepted Santer et al. comment could appear alongside it."

Osborn goes on to write that McGregor also intends to "correct the scientific record" and to identify in advance "reviewers who are both suitable and available," perhaps including "someone on the email list you've been using." Given the bias of Osborn and McGregor as expressed in the emails, it is clear that a "suitable" reviewer is someone sympathetic to Santer's views. Santer doesn't express surprise over this special treatment, implying that this was business as usual, and sets forth conditions ensuring that he will have the "last word": "1) Our paper should be regarded as an independent contribution, not as a comment on Douglass et al. . . . 2) If IJC agrees to 1), then Douglass et al. should have the opportunity to respond to our contribution, and we should be given the chance to reply. Any response and reply should be published side-by-side, in the same issue of IJC."

The Douglass et al. authors were never informed of this process, which specifically addresses our paper, nor were we contacted for an explanation on any point raised in these negotiations. Osborn instructed others on the email list to keep it a secret, writing "the only thing I didn't want to make more generally known was the suggestion that print publication of Douglass et al. might be delayed . . . all other aspects of this discussion are unrestricted. . . ."

Many interesting emails followed. One is from Santer saying he does "NOT" want to "show the most recent radiosonde [balloon] results" from Hadley Centre and Steve Sherwood's IUK (Iterative Universal Kriging) project—in other words, withholding data that were available at the time. The reason is likely that these two datasets, extended out in time, provide even stronger evidence in favor of Douglass et al. The final paper cuts off these datasets in 1999.

The conspirators at one point had difficulty hiding their work. A paper appeared in May 2008 in *Nature Geosciences* referencing the as-yet-unpublished paper by Santer et al. Douglass notices it and asks to see it . . . a critique he didn't know had been written of his paper that had yet to be published! Santer replies, "I see no conceivable reason why I should now send you an advance copy of my IJoC paper." This, despite the fact that Santer had been a reviewer of Douglass et al. when it had been submitted earlier to a different publication, so he had been in possession of the material (only slightly changed) for at least a year. Additionally, Santer had received a copy of the Douglass et al. page proofs about a week before it even appeared online.

In their *American Thinker* essay reporting this series of emails, Douglass and Christy say "we will let the reader judge whether this team effort, revealed in dozens of e-mails and taking nearly a year, involves inappropriate behavior." I do not have to be so restrained. Science depends on full publication of data and methods, replication of results, and open debate. Collaboration between authors and an editor to silence one side in a scientific debate is an egregious violation of professional ethics, as is using confidential information and withholding data. These men should have been punished. At a minimum, they should no longer be allowed to publish in the scientific literature.

In the Courtroom

In 2010, Virginia's newly elected Attorney General Kenneth Cuccinelli, following a Virginia law titled the Fraud Against Taxpayers Act, issued a civil investigative demand on the University of Virginia for Mann's emails, work notes, and other documentation. The university, a state-supported institution, resisted this demand, citing academic freedom and similar excuses. I am quite disappointed by my university's opposition to releasing those emails, which could clear up the mystery of "Mike's *Nature* trick" and reveal hidden data. I am told that no objection was raised by the university when Greenpeace requested the emails of skeptical faculty—including mine—under the Freedom of Information Act (FOIA). So much for the university's "principled defense" of academic freedom.

Virginia's Supreme Court turned down AG Cuccinelli's demand based on a technicality in the interpretation of the Virginia law. But the American Tradition Institute (now called the Energy & Environment Legal Institute) took up the effort using the FOIA. Their chance for success looked good— particularly since the university admitted it held some 12,000 emails (previously claimed to have been deleted) and had already released them to Michael Mann, even though he was no longer a faculty member. According to a Virginia Attorney General's opinion from 1983, once a public body disseminates any record, "those records lose the exemption accorded by" the FOIA. Federal case law appears to be even clearer, saying "selective disclosure . . . is offensive to the purposes underlying the FOIA and intolerable as a matter of policy." Nevertheless, in 2014, the court rejected the Institute's final appeal.

Meanwhile, a new angle developed in Vancouver, BC, in 2011 when Canadian climatologist Tim Ball jokingly wrote that "Mann should not be at Penn State but in a State Pen[itentiary]." Mann then sued Ball for libel. But this left Mann open for the pretrial discovery process, including a deposition under oath. Would he finally be compelled to share data he had kept hidden for two decades? Regrettably, no. Mann repeatedly asked the court for extensions in the case until finally, in 2019 (eight years later!), the presiding judge dismissed the case and ordered Mann to pay Ball's legal expenses. We can be happy for Ball that he won the case, but it's too bad Mann wasn't finally compelled to show his data.

* * *

The Climategate emails document a conspiracy among a clique of British and US climate scientists to control what goes into IPCC reports and to keep contrary views by skeptics from being published in recognized science journals. The scandal may have played a decisive role in shaking the public's faith in the climate science of the IPCC. They revealed key IPCC scientists were hiding their raw temperature data and the methodology of their selection and adjustments, conspiring to delete incriminating emails, and undermining the peer-review system to make it difficult for skeptical scientists to publish their work in scientific journals. In the process, this small clique damaged the whole science enterprise. For that, there can be no forgiveness.

PART TWO

Cold Science

7

What Science Really Says

SO WHERE DOES the scientific debate stand today? To answer that question, in 2003 I convened a group of scientists called Team B to challenge the alarmist reports of the IPCC. In 2007, we renamed it the Nongovernmental International Panel on Climate Change (NIPCC). Hundreds of scientists from around the world participate in NIPCC and have produced a series of volumes in a series titled *Climate Change Reconsidered*. The Chinese Academy of Sciences was so impressed by NIPCC that it translated parts of the first two volumes in the *Climate Change Reconsidered* series into Mandarin and published them as a 329-page book in 2013. A list of all books and policy reports produced by NIPCC appears in Box 4 at the end of this chapter.

While focusing on errors in the surface temperature record, gaps between observed temperatures and climate, and other somewhat technical issues, we also shall put forward some basic statements that accurately describe what scientists today know and don't know about the Earth's climate and possible human impacts on it.

The Temperature Record

Greenhouse warming has been with us for 4.5 billion years, throughout the history of the Earth. An Earth without infrared-absorbing gases in its atmosphere would be quite cold. We can calculate the surface temperature by balancing the incoming solar radiation with the outgoing heat radiation from the surface. If we assume that the Earth (with atmosphere) has an "albedo" of 0.3—i.e., reflects 30 percent of the incoming solar radiation back out into

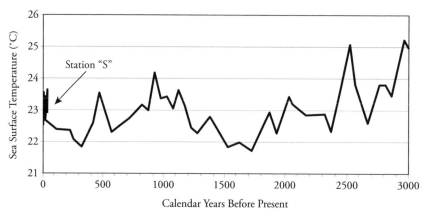

Source: Keigwin 1996.

Figure 7. Temperature variations of the past 3,000 years (during recorded history), as determined from ocean sediment studies in the North Atlantic. Note the rapid variations, as well as much warmer temperatures 1,000 and 2,500 years ago.

space—and that the IR emissivity of the surface is 1.00, the equilibrium temperature would be −18°C, well below the freezing point of water. The average temperature now is 15°C. The difference, about 33°C, can be ascribed to the natural greenhouse effect, produced mainly by water vapor and carbon dioxide (CO_2) in the atmosphere.

Earth's surface and tropospheric temperatures also have been variable since the dawn of time. Variations appear in the temperature record of the past 3,000 years as determined from ocean sediment studies in the North Atlantic (see Figure 7) and during the past 10,000 years as shown in studies of Greenland ice cores (see Figure 8). Note the rapid variations, as well as much warmer temperatures 1,000 and 2,500 years ago when the atmospheric CO_2 concentration was only about 200 ppm, rather than 280 ppm, the preindustrial value during most of the Holocene.

America's preeminent climate "oracle," Al Gore, made an apocalyptic movie in 2006, *An Inconvenient Truth*. Early in 2006, when promoting the movie, he predicted that unless we took "drastic measures" to reduce GHGs, the world would reach a "point of no return" in a mere ten years. He called it a "true planetary emergency." The basis of Gore's prediction was a correlation of

a. Greenland GISP2 oxygen isotope curve for the past 10,000 years

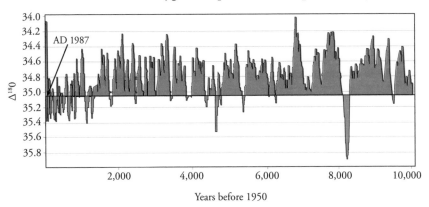

b. Greenland GISP2 oxygen isotope curve for the past 5,000 years

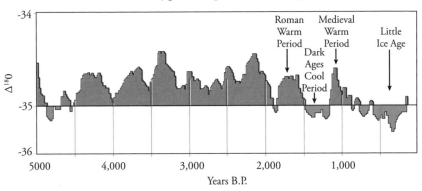

Source: Alley 2000.

Figure 8. Temperatures from Greenland ice cores. The vertical axis is δ ^{18}O, which is a temperature proxy. Horizontal scale for (a) is 10,000 years before 1950 and for (b) is past 5,000 years. The red areas represent temperatures warmer than present (1950). Blue areas are cooler times. Note the abrupt, short-term cooling 8,200 years ago, shown in (a), and cooling from about AD 1500 to 1950, shown in (b).

several sudden temperature rises and CO_2 increases during the recent ice age, as judged from analysis of Antarctic ice cores. He then erroneously concluded that this correlation proved that CO_2 *caused* twentieth-century warming.

To Gore's great embarrassment, it was revealed that the increase in CO_2 actually *followed* the temperature increase by about 600 to 800 years. (Dennis Avery and I told the story of that discovery in our book titled *Unstoppable Global Warming: Every 1,500 Years.*) The mechanism is really quite simple: When the ocean warms, it releases much of its dissolved CO_2, similar to warming soda pop or a bottle of champagne releasing CO_2 bubbles. Even a nonscientist understands that the cause must always precede the effect. Suddenly, the "smoking gun" that Al Gore relied on collapsed in a heap.

Laurie David, the producer of Al Gore's movie and author of popular cookbooks, and Cambria Gordon, "a former award-winning advertising copywriter," coauthored a widely acclaimed children's book in 2007 titled *The Down-to-Earth Guide to Global Warming.* According to Amazon.com, the book is "irreverent and entertaining" and "filled with fact about global warming and its disastrous consequences." It features a figure showing "CO_2 concentration in the atmosphere" and "climate temperature" during the past 650,000 years and claims it demonstrates "the link between greenhouse-gas pollution and global warming." But to make CO_2 appear to lead rather than follow changes in temperature, the authors *switched the labels* of the two lines in the figure. The line showing CO_2 concentration is labeled temperature and the line showing temperature is labeled CO_2 concentration. This is an elegant solution to an embarrassing fact, but it is statistical fraud, something that would rightly end an academic's career. When confronted with the falsehood immediately after the book was published by a group called the Science & Public Policy Institute, David dismissed it as a "minor error" (Ferguson 2007).

Both paleoclimatological and historical data confirm the existence of a warm period, the MWP (950–1250 AD), and a cold period, the LIA (1400–1850 AD), with temperature departures about ± 1°C (1.8°F) from the mean of the Holocene. Since we have no explanation for these oscillations—and even larger ones in the past (Keigwin 1996)—we must accept the possibility that they will continue. Even the IPCC in its First Assessment Report admitted that "some of the global warming since 1850 could be a recovery from the Little Ice Age rather than a direct result of human activities" (IPCC AR1, WGI, 203).

What We Think We Know

Some basic facts about Earth's climate are not seriously disputed in the science community. Climate change is historical fact and occurs on many different time scales. Paleoclimatologists, using the disciplines of geology, astrophysics, isotope chemistry, oceanography, biology, and more, have reached certain tentative conclusions. Basically, the cause is related to the time scale under consideration:

- Hundreds of thousands to million years—variations in galactic cosmic rays
- Ten thousand to one hundred thousand years—changes in Earth orbit and motion
- Decades to centuries—natural internal variability and solar variability

It is a basic fact that climate change has occurred on a scale larger than what was observed in the twentieth century, before there could have been a human role. If the climate is indeed moving out of the LIA and possibly into another climatic optimum, as many paleoclimatologists believe, the roughly 0.5°C warming of the past 130 years of observational record is evidence of neither greenhouse warming nor anthropogenic effect. We could even look forward to an additional warming of about 1°C over the next few centuries regardless of what humans do.

Second, most scientists agree global temperatures rose during the early part of the twentieth century, up to about 1940. It then cooled until about 1975, raising widespread fears of a coming Ice Age. A sudden rise of nearly 0.2°C occurred between 1976 and 1978, linked to a shift in ocean circulation. Upward shifts in temperature also occurred in 1997–98 and 2007–8, coinciding with changes in ocean currents. There is no known connection between ocean currents and the concentration of CO_2 in the atmosphere.

Third, there is no dispute that levels of greenhouse gases—carbon dioxide (CO_2), methane (CH_4), nitrous oxide (N_2O), chlorofluorocarbons (CFCs), etc.—in the atmosphere have increased as a result of human activities. Human CO_2 emissions have been increasing at about 0.5 percent per year, mostly as a result of fossil-fuel combustion related directly to energy generation. The current level of atmospheric CO2, about 412 ppm (parts per million) in 2019, is about 130 ppm (46 percent) higher than the preindustrial level of 280 ppm. The sources of methane, whose concentration has doubled in the past one hundred years, are more varied: in addition to natural sources such as swamps and wetlands, human sources include fossil-fuel operations and landfills as well as cattle raising and rice growing.

Fourth, water vapor is responsible for most of the greenhouse effect. Without the presence of naturally occurring atmospheric GHGs like CO2 and especially water vapor, our planet would be a frozen wasteland with oceans covered by ice. Ice has a very high reflecting power (albedo) in the visible region of the spectrum, while open ocean water reflects somewhat less than 10 percent of the incoming solar radiation. Once the oceans freeze, they could never recover. One of the mysteries of the Earth's early history is how the oceans managed to stay liquid at a time when solar radiation was less than 80 percent of its present value, called the Faint Young Sun Paradox (Caldeira and Kasting 1992).

Finally, based on solid evidence, agriculturists concur that the ongoing increase in atmospheric CO2 speeds up plant growth. Empirical evidence indicates that a modest climate warming, from whatever cause, would benefit plant and animal life in nearly all parts of the globe. A "greening of the Earth" is already being seen from satellites (Zhu et al. 2016). The only disagreement on this point is over *how much* plants and animals would benefit and at what temperature those benefits stop accruing.

What We Know We Don't Know

What we think we know about climate change is dwarfed by what we know we don't know. To begin, some scientists dispute the scientific validity of a single global temperature estimate inferred from measurements from random parts of the world and interpreted by models. For example, Christopher Essex and colleagues conclude that "there is no physically meaningful global temperature for the Earth in the context of the issue of global warming" (Essex, McKitrick and Andresen 2007, 1). According to Thomas Peterson and Russell Vose, "there are over 100 different ways in which daily mean temperature has

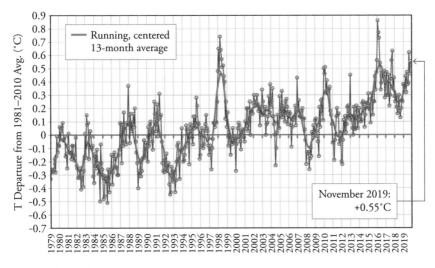

Source: Spencer 2019.

Figure 9. Satellite-based temperature record, 1979–2019, according to the Earth System Science Center at the University of Alabama–Huntsville. Blue circles and lines show monthly global-average temperature anomalies for the lower troposphere, while the red line is the running, centered 13-month average.

been calculated by meteorologists" (Peterson and Vose 1997, 2841). Which one is "right" is impossible to know.

There is an important disagreement about the temperature record since 1979, with data from satellites and radiosondes (carried by weather balloons) showing only a slight warming of approximately 0.1°C per decade (Christy et al. 2018), whereas surface thermometers show a warming trend about three times as great. (See Figures 9 and 10.) The surface station record (HadCRUT) has been severely criticized for inaccuracies and methodological errors (Fall et al. 2011; McLean 2018), a subject addressed in some detail in Chapter 8. It is important to note that atmospheric data taken with balloon-borne radiosondes independently confirm the satellite data.

The use of computer models to estimate climate sensitivity to GHGs and then to forecast future global temperatures that would result from various emissions scenarios is also a source of controversy. Satellites, weather balloons,

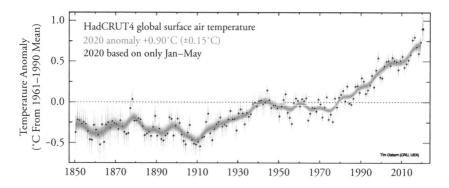

Source: HadCRUT4 temperature dataset on the CRU website, https://crudata. uea.ac.uk/~timo/diag/tempdiag.htm.

Figure 10. Monthly global temperature anomalies, 1850–2020 according to the Hadley Centre of the UK Met Office and University of East Anglia Climatic Research Unit (CRU). Black/gray dots are the annual average temperature anomalies, vertical lines are their estimated uncertainty. Decadally-smoothed values are shown in blue, with shading to indicate the 95 percent confidence interval at the decadal timescale. The most recent year (2020) is shown in red.

and even the unreliable surface temperature station record all show much lower warming trends than what computer models predict, as is readily apparent in Figure 11. (The methodology behind this figure and John Christy's interpretation of it are presented in Chapter 9.)

Prominent computer modelers have admitted to "tuning" or "tweaking" their models to arrive at forecasts they believe their funders and colleagues want to see (Voosen 2016; Hourdin et al. 2017). The key question is why rising levels of GHGs in the atmosphere are not causing a global warming in accord with the expectations from current climate models. Only if these GCMs are validated through observations can one rely on their forecasts of future warming.

The question of attribution—whether changes in global temperature or other climate elements can be attributed to GHG concentrations in the atmosphere and, more specifically, to CO_2 concentrations—also remains an issue of controversy. Historic data show CO_2 concentrations in past geological

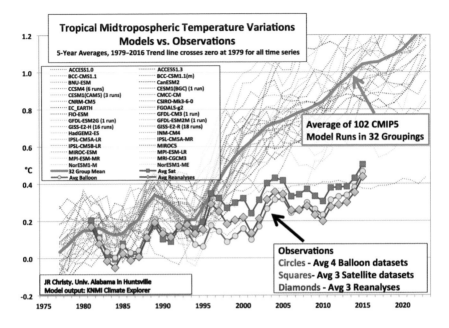

Source: Christy 2017.

Figure 11. Climate model forecasts versus observations, 1979–2016. Five-year averaged values of annual mean (1979–2016) tropical bulk T_{MT} as depicted by the average of 102 IPCC CMIP5 climate models (red) in 32 institutional groups (dotted lines). The 1979–2016 linear trend of all-time series intersects at zero in 1979. Observations are displayed with symbols: green circles—average of four balloon datasets; blue squares—three satellite datasets; and purple diamonds—three reanalyses. The last observational point at 2015 is the average of 2013–2016 only, while all other points are centered, five-year averages.

periods up to twenty times greater than the present value—without harming the climate system. There seem to be no obvious connections: While the large fluctuations of the ice ages of the past 2 million years arose after CO_2 levels had fallen to near-present levels, there was a period of widespread glaciation during the Ordovician (440 million years ago) when CO_2 levels were fifteen times the present value. Ice core data also show climate fluctuations were much greater during the lower CO_2 levels of the most recent ice age than at the higher CO_2 levels of the present warm interglacial (Holocene) period of the past 10,000 years. Does this result suggest that higher CO_2 levels promote more climate stability and therefore present less "danger to the climate system"?

According to the IPCC, "It is extremely likely that human influence on climate caused more than half of the observed increase in global average surface temperature from 1951–2010" (IPCC; AR5; WGI; Summary for Policymakers 2013, 17). But GHG concentrations have already gone halfway toward a CO_2-equivalent doubling, mainly in the past fifty years, while the climate record shows no commensurate warming since 1940. Surveys show most scientists and meteorologists believe natural causes, not human influence, caused the majority of the temperature rise, and even those who contributed to AR5 admit to lacking high confidence in such a statement (Kummer 2015; Maibach et al. 2017). This does not argue that AGW is absent, but that it is simply too small to be detectable, and much less than calculated from GCMs.

Experts agree that human CO_2 emissions are likely to increase as Western nations continue to grow in population and prosperity and developing countries use greater amounts of fossil fuels to lift their populations up from poverty. However, considerable uncertainty exists regarding how quickly and how high such emissions will rise and consequently their concentration in the atmosphere. Predictions of future emissions (what the IPCC now calls "representative concentration pathways" or RCPs) depend on assumptions about population and economic growth rates and technological and political changes occurring decades and even centuries in the future. "Futurists" have shown little skill in making predictions about these important variables reaching out longer than just a few years (Armstrong and Green 2018).

New sources of emissions and "sinks" are reported frequently in scientific journals (e.g., Bastin et al. 2017; Wylie 2013), leading to uncertainty about the size of Earth's carbon reservoirs (atmosphere, oceans, biosphere, and lithosphere) and exchange rates among these reservoirs. This is important because the human contribution of CO_2 to the atmosphere that is thought to remain there for more than a short period (before being absorbed by the oceans and biosphere) is very small relative to natural processes, just 0.53 percent of the carbon entering the atmosphere each year (IPCC, AR5, WGI, 2013, 471).

Even small errors in the measurement of poorly understood exchange processes exceed the entire human contribution by orders of magnitude.

The climate record gives little guidance as to what constitutes the "right" or "best" atmospheric levels of CO_2. The weather of the 1940s, 1990s, or 2010s may or may not be the "right" or "best" global temperature for humans or for the natural world, so attempting to return to it or preserve it does not mean making the planet's climate better or safer. Humanity thrived during warmer periods of history and suffered during cold spells (Singer and Avery 2008). Moderate warming may produce more benefits than harms. On this important matter, much or even most of the scientific community disagrees with the IPCC. The IPCC's reports are filled with every possible negative effect of a slightly warming planet that can be imagined: floods, droughts, more frequent severe weather events, famines, diseases, extinction of species, forced migration, and even more wars and other civil conflicts. But conspicuously absent from the IPCC's reports is any admission of the offsetting benefits of a warmer world with higher levels of CO_2 in its atmosphere. Those benefits are discussed in Chapter 13.

Finally, a great deal of effort has gone into finding trends in weather and wildlife data that may support the IPCC's predictions of catastrophic effects in the years ahead. Such claims invariably rely on cherry-picking years, relying on regional rather than global databases, or artful reinventions of how weather events are defined. Here are what scientists really know about trends in stormy weather, heat waves, droughts, and coral health. (Sea level rise, which is a particularly controversial issue, is discussed in Chapter 11.)

Severe Storms Are Not Increasing

Searching the climate record for past weather and storm patterns can reveal whether the warming of the twentieth and early twenty-first centuries was accompanied by more severe weather, such as thunderstorms and hurricanes. The limited availability of observational data and known deficiencies of climate models used to simulate storm-producing processes make generalizations difficult (Zhang et al. 2017). However, data suggest the frequency of thunderstorms and related activity *declined* during the second half of the

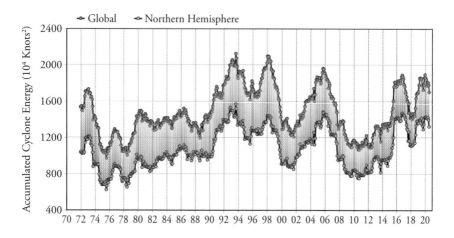

Source: Maue 2020.

Figure 12. Cyclonic energy, globally and Northern Hemisphere, from 1970 through August 2020. Last four decades of global and Northern Hemisphere Accumulated Cyclone Energy (ACE): twenty-four-month running sums. Note that the year indicated represents the value of ACE through the previous twenty-four months for the Northern Hemisphere (bottom line/gray boxes) and the entire globe (top line/blue boxes). The area in between represents the Southern Hemisphere total ACE.

twentieth century and more recently in Australia, Canada, China, Europe, New Zealand, and the United States (NIPCC 2019, 210–16). Hurricanes globally and in the Northern Hemisphere specifically show no long-term trend in frequency and intensity since 1970 when satellite monitoring began (Pielke et al. 2005, Klotzbach et al. 2018). (See Figure 12.)

Heat Waves Are Not Becoming More Common

According to the IPCC's Fifth Assessment Report, "It is virtually certain that there will be more frequent hot and fewer cold temperature extremes over most land areas on daily and seasonal timescales as global mean temperatures increase. It is very likely that heat waves will occur with a higher frequency and duration" (IPCC, AR5, WGI, 2013, 20). But familiarity with the climate

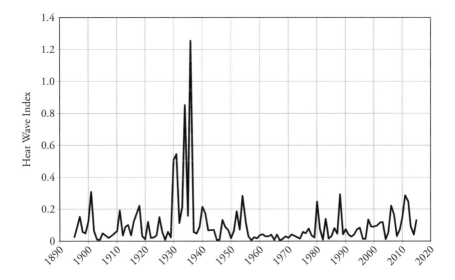

Source: EPA 2016.

Figure 13. US heat wave index, 1895–2015. These data cover the contiguous forty-eight states. An index value of 0.2 could mean that 20 percent of the country experienced one heat wave, 10 percent of the country experienced two heat waves, or some other combination of frequency and area resulted in this value.

record reveals this fear is unwarranted. Extensive investigation of historical records and proxy data has found many examples of absolute temperature or variability of temperature exceeding observational data from the twentieth and early twenty-first centuries (Bohm 2012; Rusticucci 2012; Christy 2012). Figure 13 shows the absence of any increase in the area of the United States experiencing heat waves (defined by the US EPA as a period lasting at least four days with an average temperature that would be expected to occur only once every ten years, based on the historical record) between 1895 and 2015.

Droughts Are Not Becoming More Common

Higher surface temperatures are said to result in more frequent, severe, and longer-lasting droughts. Even the IPCC expresses doubt that this has occurred

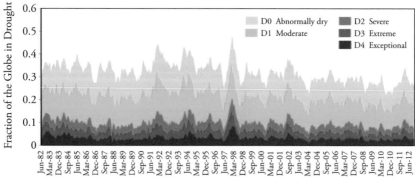

Source: Hao et al. 2014.

Figure 14. Global areal extent of five levels of drought for 1982–
2012. Fraction of the global land in D0 (abnormally
dry), D1 (moderate), D2 (severe), D3 (extreme), and D4
(exceptional) drought condition. Data: Standardized
Precipitation Index data derived from MERRA-Land.

in recent decades. The authors of AR5 write "compelling arguments both for
and against significant increase in the land area affected by drought and/or
dryness since the mid-20th century have resulted in a *low confidence* assessment of observed and attributable large-scale trends" and *"high confidence*
that proxy information provides evidence of droughts of greater magnitude
and longer duration than observed during the 20th century in many regions"
(IPCC, AR5, 2013, 112).

The historical record is replete with accounts of megadroughts lasting
for several decades to centuries that occurred during the MWP, dwarfing
modern-day droughts (e.g., Seager et al. 2007; Cook et al. 2010). Atmospheric
CO_2 concentrations were more than 100 ppm lower during the MWP than
they are today. The clear implication is that natural processes operating during the MWP were responsible for droughts that were much more frequent
and lasted much longer than those observed in the twentieth and twenty-first
centuries.

Looking at more recent trends, Hao and colleagues (2014) found the
global areal extent of drought fell from 1982 to 2012 across all five levels used
to rank drought conditions. A figure illustrating their findings is reproduced
as Figure 14.

Global Warming Is Not Harming Coral Reefs

Laboratory experiments have demonstrated that changes in ocean water chemistry can lead to reductions in the calcium carbonate saturation state of seawater, which lowers the water's pH level. This has led to predictions that elevated levels of atmospheric CO_2 may reduce rates of coral calcification, possibly leading to slower-growing—and, therefore, weaker—coral skeletons, and in some cases even death (Barker and Ridgwell 2012). In the same way, changes in temperature and salinity might also have detrimental effects on coral.

Such claims are based on faulty model predictions. Hugo Loaiciga reports that "a doubling of CO_2 from 380 ppm to 760 ppm increases the seawater acidity [lowers its pH] approximately 0.19 pH units," an amount far below natural variability (Loaiciga 2006, 1). He concludes that "on a global scale and over the time scales considered (hundreds of years), there would not be accentuated changes in either seawater salinity or acidity from the rising concentration of atmospheric CO_2" (Loaiciga 2006, 3).

Similarly, Pieter Tans estimates the decline in oceanic pH by the year 2100 is likely to be only about half of that projected by the IPCC and that this drop will begin to be ameliorated shortly after 2100, gradually returning oceanic pH to present-day values beyond AD 2500 (Tans 2009).

Predictions that global warming would harm corals also fail to account for the fact that there is nothing unusual, unnatural, or unprecedented about recent pH values or values forecast to exist in the future. Liu et al. (2009) reconstructed the paleo-pH record of the past 7,000 years for the South China Sea, depicted in Figure 15. The two most recent pH values (shown on the far-right edge of the figure) are below most but not all values during the historical record and nowhere near the 7.0 pH boundary for acid conditions.

In the real world, living organisms find ways to meet and overcome the many challenges they face. Coral calcification in response to so-called "ocean acidification" is no exception (McCulloch et al. 2017). Bleaching is part of corals' strategy for adapting to often changing water temperatures (Hecht 2004). Cynthia Lewis and Mary Alice Coffroth of the University of Buffalo deliberately triggered bleaching in some coral colonies (Lewis and Coffroth 2004). In response, the colonies ejected 99 percent of their symbiotic algae

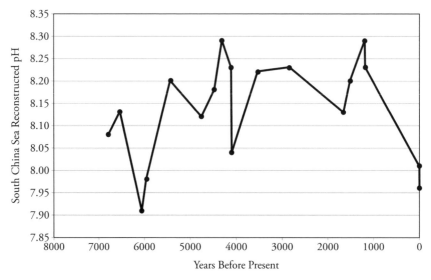

Source: Created from Table 1 of Liu et al. 2009.

Figure 15. Reconstructed pH history of the South China Sea.

friends. The researchers then exposed the bleached coral to a rare variety of algae that wasn't in the coral colonies at the beginning of the experiment. Sure enough, within a few weeks, the corals had substantially restocked their algae shelves, and about half included the new marker algae. Later, the marker variety was displaced from several of the coral colonies by more effective algae strains—indicating that the corals pick the best partners for the new conditions from the wide variety of algae floating in their part of the ocean.

Lewis and Coffroth say this is a healthy demonstration of flexibility in coral colonies. They say coral systems have the flexibility to establish new associations with algae strains from the whole environmental pool and that this is "a mechanism for resilience in the face of environmental change" (Lewis and Coffroth 2004, 7).

More recently, Jessica Carilli, Simon Donner, and Aaron Hartmann (2012) write "there is evidence that corals may adapt to better withstand heat stress via a number of mechanisms," noting "corals might acquire more thermally-resistant symbionts, or might increase their own physiological mechanisms to reduce bleaching susceptibility by producing oxidative enzymes or photoprotective compounds." They further point out that the susceptibility of a given coral or reef to bleaching depends on the thermal history of that coral or reef.

Claims that global warming is killing corals are simply wrong. Science has returned its verdict on corals and global warming: No link.

* * *

So, what does science really tell us about climate change? It's very different from what one might read in, say, the *New York Times* or even, sadly, in editorials in *Nature* and other once-prestigious science journals. We know climate change is a permanent feature of planet Earth; any human impact that might be occurring is probably too small to discern against a background of natural variability; and CO_2, so often blamed for changing the weather, is almost surely a minor player compared to natural processes. Despite all the hot talk, there is no "climate crisis" resulting from human activities and no such thing on the horizon.

In spite of the fact that science does not support the emission cuts and other policies that dominate the debate today, I do not argue for complacency. Any human-induced change in environment must be carefully monitored and evaluated. In the meantime, however, commonsense "no-regrets" policies like cost-effective energy conservation and improved efficiency are in order, rather than hasty and economically damaging actions based on insufficient science.

Box 4

NIPCC Publications

Below is a list of all books and policy reports produced by the Nongovernmental International Panel on Climate Change (NIPCC) to date. They are all available for free in pdf format at www.nipccreport.org. Print versions can be ordered at store.heartland.org or Amazon.com.

Title	Date	Pages	Lead Authors
Nature, Not Human Activity, Rules the Climate	2008	40	S. Fred Singer
Climate Change Reconsidered: The 2009 Report of the Nongovernmental International Panel on Climate Change	2009	856	Craig D. Idso, S. Fred Singer
Climate Change Reconsidered: 2011 Interim Report	2011	416	Craig D. Idso, S. Fred Singer, Robert Carter
NIPCC vs. IPCC (published by European Institute for Climate and Energy [EIKE] in English and German)	2011	28	S. Fred Singer
Climate Change Reconsidered II: Physical Science	2013	993	Craig D. Idso, Robert M. Carter, S. Fred Singer
Climate Change Reconsidered: The Report of the Nongovernmental International Panel on Climate Change (published by the Chinese Academy of Sciences in Mandarin)	2013	329	Craig D. Idso, Robert M. Carter, S. Fred Singer
Written Evidence Submitted to the Commons Select Committee of the United Kingdom Parliament	2013	7	Craig Idso, Robert M. Carter, S. Fred Singer

Scientific Critique of IPCC's 2013 "Summary for Policymakers"	2013	18	Craig Idso, Robert M. Carter, S. Fred Singer, Willie Soon
Climate Change Reconsidered II: Biological Impacts	2014	1,062	Craig D. Idso, Sherwood Idso, Robert M. Carter, S. Fred Singer
Commentary and Analysis on the Whitehead & Associates 2014 NSW Sea-Level Report	2014	44	Robert M. Carter, et al.
Why Scientists Disagree About Global Warming: The NIPCC Report on Scientific Consensus	2015 (Second edition, 2016)	110	Craig D. Idso, Robert M. Carter, S. Fred Singer
Data versus Hype: How Ten Cities Show Sea-Level Rise Is a False Crisis	2017	13	Dennis Hedke
Global Warming Surprises: Temperature Data in Dispute Can Reverse Conclusions About Human Influence on Climate	2017	6	S. Fred Singer
Climate Change Reconsidered II: Fossil Fuels	2019	784	Roger Bezdek, Craig D. Idso, David R. Legates, S. Fred Singer

8

The Unreliable Surface Temperature Record

EXPLORING SOME OF the intricacies of climate in the twentieth century can lead to surprising results that have major consequences for our understanding of climate change. One such surprise is inconsistencies in the surface temperature record and the problems in data collection that produce them.

Two Warming Periods or One?

In one of the iconic pictures of the global surface temperature of the past century, shown in Figure 16, one can discern two warming intervals, in the initial decades (1910–45) and in the final decades (1979 to present). For their proofs of human influence, the IPCC's AR4 and AR5 relied on warming observed between 1978, which followed a step increase in temperatures thought to be due to a change in the Pacific Decadal Oscillation (PDO), and 1997, the year before a super–El Niño.

The first conclusion most people draw from looking at Figure 16 is that the rate of increase in temperature during the first period, 1910–45, appears to be about the same as that during the second period, 1978–97. Indeed, Phil Jones, director of the CRU, when asked in 2010 (in writing) if the rates of global warming from 1860–80, 1910–40, and 1975–98 were identical, wrote in reply that "the warming rates for all 4 periods [he added 1975–2009] are similar and not statistically significantly different from each other" (Harrabin 2010). When asked, "Do you agree that from 1995 to the present there has been no statistically significant global warming?" Jones answered "yes."

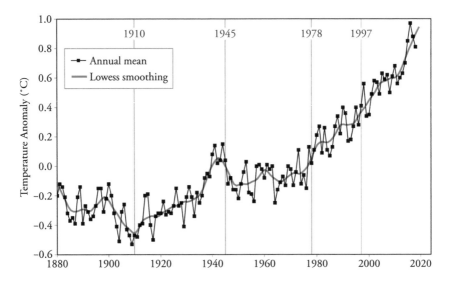

Source: NASA/GISS 2019.

Figure 16. NASA/GISS estimated global mean surface temperature anomaly from 1880 to 2019, with base period 1951–80, according to NASA's Goddard Institute for Space Studies (GISS). The solid black line is the global annual mean and the solid red curve is the five-year lowess smoothing. The blue uncertainty bars (95 percent confidence limit) for three years account only for incomplete spatial sampling. Note the absence of warming from about 1945 to 1978.

As damaging as Jones's admission was to the IPCC's claim that the warming in the past century was "unprecedented," an even more important story can be told about these two trends. It is my contention that although these two trends *look similar*, they are really quite different. The warming from 1910 to 1945 was real: It is confirmed by thermometer records as well as proxy data. The warming during the later period, from 1978 to 1997, *is almost entirely fake*, an instrumental artifact found only in the HadCRUT database of surface observations.

We start by observing that datasets more reliable than HadCRUT do not show the rise in temperature during the period 1979–97 that appears in Figure 16. This is summarized in Box 5.

Box 5

Lack of Evidence for Rising Temperatures from 1979 to 1997

LAND SURFACE	Global (IPCC, HadCRUT)	~+0.5-0.7°C
	US (GISS, BEST)	~zero
OCEAN	Sea surface temperature	~zero
	Hadley NMAT	~zero
ATMOSPHERE	Satellite MSU (1979–97)	~zero
	Hadley Radiosondes (1979–97)	~zero
PROXIES	Mostly land surface temperatures	~zero

IPCC = Intergovernmental Panel on Climate Change; HadCRUT = Hadley Centre of the UK Met Office and Climatic Research Unit (CRU) of the University of East Anglia; NMAT = Night Time Marine Air Temperatures; GISS = Goddard Institute for Space Studies; BEST = Berkeley (CA) Earth System Temperatures; MSU = Microwave Sounding Unit

Specifically, during the period from 1979–97,

- the surface record for the *forty-eight contiguous US states* shows a much lower trend than the global trend (Karl and Jones 1989); this is significant because the US system of weather stations is generally regarded as being better than those of much of the rest of the world (though still flawed);

- the trend of global sea surface temperature (SST) is much less, with 1995 temperature values nearly equal to those of 1942 (Gouretski et al. 2012);

- the trend of nighttime marine air temperatures (NMAT), measured with thermometers on ship decks, is approximately zero (Kent et al. 2013), see Figure 17;

- balloon-borne radiosondes and satellites with microwave sounding units (MSU) aboard show little or no warming (Spencer 2018), see Figure 9 in Chapter 7;

- proxy data show near-zero trends for this interval, whether from tree

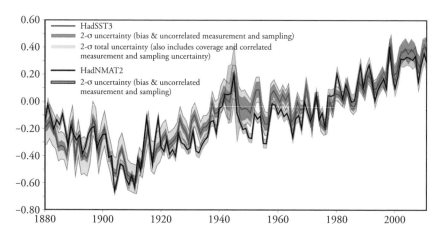

Source: Kent et al. 2013.

Figure 17. Global annual average nighttime marine air temperature (HadNMAT2) and sea surface temperature (HadSST3) median anomalies (°C, relative to 1961–90) and their estimated uncertainties, 1880–2010. Note that while there is a rise from 1979–97, it is much less than what is shown in Figure 16 and the period ends below the temperature in 1945.

rings or ice cores (Jacoby, D'Arrigo, and Davaajamts 1996; McIntyre and McKitrick 2003; Dahl-Jensen et al. 1998).

By contrast, the early warming (1910–45) is supported by many proxy data including temperatures derived from tree rings and ice cores. Note that I choose for this analysis the period 1979 to 1997, rather than the longer surface temperature record (beginning in 1880) or the satellite record (beginning in 1979) which continue to the present, because this period is between two step-wise increases in global temperature that cannot be explained by the greenhouse theory and so confound the data with natural variation. If the greenhouse theory is correct, warming should have been observed during this period. It was not.

In 2000, the National Research Council tried to account for the disparity between surface thermometers showing a warming trend (those databases that did not control for the effects of ENSO events) and weather satellite and (independent) weather balloon/radiosonde observations that show no appreciable warming of the lower atmosphere since 1979. The study concluded that both were correct; the satellites were showing no warming in the lower troposphere while the surface thermometers were reporting rising temperatures on the surface (National Research Council 2000). The executive summary of the report misrepresented this finding as endorsing the accuracy of the surface temperature record, but climate models predict warming *in the troposphere* and not at the surface, so it is clear which database ought to be used when testing the validity of climate models.

Corrupt Data and Changes in Instrumentation

The IPCC admits its temperature reconstructions are highly uncertain. The authors of the Fifth Assessment Report (2013) write,

> The uncertainty in observational records encompasses instrumental/ recording errors, effects of representation (e.g., exposure, observing frequency or timing), as well as effects due to physical changes in the instrumentation (such as station relocations or new satellites). All further processing steps (transmission, storage, gridding, interpolating, averaging) also have their own particular uncertainties. Because there is no unique, unambiguous, way to identify and account for non-climatic artefacts in the vast majority of records, there must be a degree of uncertainty as to how the climate system has changed (IPCC, AR5, WGI, 165).

Recall from the previous chapter that there are over one hundred different ways in which daily mean temperature has been calculated by meteorologists. Vastly increased computer power has probably multiplied that number. Efforts to manipulate and "homogenize" divergent datasets, fill in missing data, remove outliers, and compensate for changes in sampling technology are all opportunities for subjective or poor decision-making.

An audit of the HadCRUT4 dataset conducted by John McLean, an Australian scientist, found "more than 70 issues of concern," including failure to check source data for errors, resulting in "obvious errors in observation station metadata and temperature data" (McLean 2018, 88). He found the dataset "has been incorrectly adjusted in a way that exaggerates warming" (i).

Evidence of corrupted data appeared in emails from a programmer responsible for maintaining and correcting errors in the HadCRUT climate data between 2006 and 2009. His comments included "Wherever I look, there are data files, no info about what they are other than their names. And that's useless . . . "; "It's botch after botch after botch"; "Am I the first person to attempt to get the CRU databases in working order?!!" ; and "I'm hitting yet another problem that's based on the hopeless state of our databases. There is no uniform data integrity, it's just a catalogue of issues that continues to grow as they're found" (Goldstein 2009).

In 2009, in response to an academic's request for the HadCRUT dataset, Phil Jones, director of the CRU at the University of East Anglia, admitted, "We, therefore, do not hold the original raw data but only the value-added (i.e., quality controlled and homogenized) data" (Michaels 2009). As Patrick Michaels commented at the time, "If there are no data, there's no science."

Studies of the siting of weather stations in the United States, thought to have the best network of such stations in the world, find extensive violations of siting rules leading to contamination by urban heat islands (Pielke et al. 2007a, 2007b). The IPCC claims to control for heat island effects, but researchers have found its adjustments are too small (e.g., Kalnay and Cai 2003; McKitrick and Michaels 2007; Soon, Connolly, and Connolly 2015).

Changes in instrumentation also account for inaccuracies in the surface temperature record. As seen in Figure 18, the number of land temperature stations began to decrease during the 1970s and then dropped suddenly in the early 1990s. This had a large effect on the geographical distribution of stations; for example, many stations in cold weather areas of the Soviet Union were closed, likely causing some spurious warming trends to be reported. Many of the stations that remained open were near airports, as is also shown in Figure 18. The fraction of stations located at airports rose from ~35 percent to ~80 percent, producing even more spurious temperature increases due to the thermal properties of runways and buildings and heat generated by aircraft and terminals.

Sea surface temperature (SST) data also have been corrupted by changes in instrumentation. Data taken from floating drifter buoys, aggregated by weather satellites in batches of three or more for preponderance decision calibration, increased from 0 percent of all SST data to 60 percent between 1980

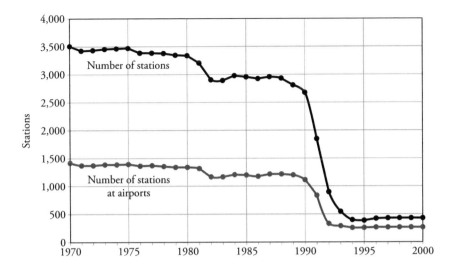

Source: Data from National Centers for Environmental Information (NOAA), "Global Historical Climatology Network (GHCN)" website, ftp: //ftp. ncdc.noaa.gov/pub/data/ghcn/v2.

Figure 18. Number of weather stations in the Global Historical Climatology Network (GHCN) from 1970 to 2000. Upper curve shows total number of stations, lower curve is number of stations located at airports. The percentage of stations located at airports rose from ~35 percent to ~80 percent during this period.

and 2000 as they replaced sampling with water buckets, while temperatures obtained from engine room inlet (ERI) water remained about the same, as shown in Figure 19.

The change in instrumentation documented in Figure 19 has a major impact on SST data collected because engine inlet water is drawn from lower (cooler) ocean layers while floating drifter buoys are heated directly by the Sun and sample the warmer surface level of water, as indicated in the diagram in Figure 20. Moreover, buoy data are global while bucket and inlet temperatures are confined to mostly commercial shipping routes. Nor do we know the ocean depths that buckets sample, and inlet depths depend on ship type and degree of loading, which changes over time.

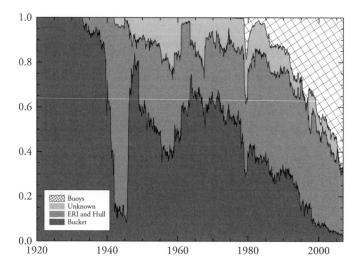

Source: Kennedy et al. 2011, figure 2.

Figure 19. Fractional contribution to the monthly average SST
from different measurement methods and platforms,
1920–2006: buckets (dark gray), ERI and hull contact
(medium gray), unknown (light gray), and buoys (cross-
hatched).

Disentangling this mess requires data details that are not available. About
all we might demonstrate is the possibility of a distinct diurnal variation in
the buoy temperatures. The transition from reliance on one type of data to
another leads to a spurious rise in SST. We have, however, satellite data for
the lower atmosphere over both ocean and land; they show little difference,
so we can assume that bias in both land data and ocean data contribute about
equally to the fictitious surface trend reported for 1978 to 1997.

* * *

The most basic and logically the most important data in the climate change
debate are past and present temperatures. How many people know this key
information is often missing, uncertain, or clearly manipulated? How many

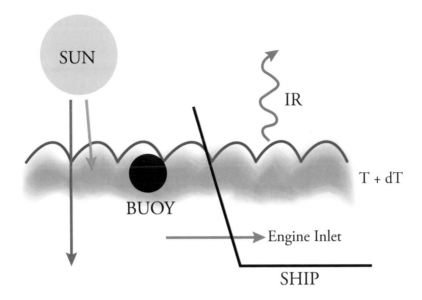

Figure 20. Why buoys record warmer temperatures than thermostats located near engine inlets. Floating buoys measure temperatures in the solar-heated layer of water near the surface, while instruments located near the inlets for engine-cooling water measure temperatures from cooler water below the surface.

people know the only global and precise measurement of global temperatures since 1979 shows almost no warming, just a tenth of a degree Celsius per decade, and most or even all of that can be explained by natural forces? Why are those data absent from reports issued by the IPCC and all the government agencies and environmental groups claiming a "climate crisis" exists?

How significant is this? Very, indeed! The absence of a warming trend removes all of the IPCC's evidence for AGW. Obviously, if there is no warming trend, these demonstrations fail and, with them, the entire case for action to slow or stop climate change from occurring.

9

The Gap Between Observed Temperatures and Climate Models

DUE TO THE shortcomings of the HadCRUT4 database and other temperature records based on surface-based weather stations, the only reliable global temperature record is the one derived from satellite-based readings of lower-atmosphere temperatures taken since 1979. When that forty-year record is used to test the accuracy of GCMs that purport to show the impact of human activity on Earth's climate, the models invariably fail, revealing that man-made CO_2 has little or even no influence on global temperatures.

No Warming Trend

After taking into account inconsistencies in the global temperature record, it is clear there has been little global warming since 1998 and even earlier in many areas of the world. The IPCC has admitted there is uncertainty regarding the measurement of global temperatures. In its latest report (AR5), released in 2013, the IPCC admitted that the global mean average temperature stopped rising for the fifteen-year period from 1998 to 2012 (AR5 2013, 5). After the super–El Niño of 1997–98, the "pause," as it is sometimes called, or hiatus continued to the present day, as shown by satellite data plotted in Figure 9 in Chapter 7 and reproduced here as Figure 21 (Spencer 2019).

John Christy's and R. T. McNider's estimate, published in the *Asia-Pacific Journal of Atmospheric Sciences* in 2017, of a warming of approximately 0.10°C per decade since 1979, when the satellite record begins, includes the step increase in global temperature caused by the super–El Niño in 1997–98 and the strong El Niño of 2015–16, which cannot be explained by greenhouse warming.

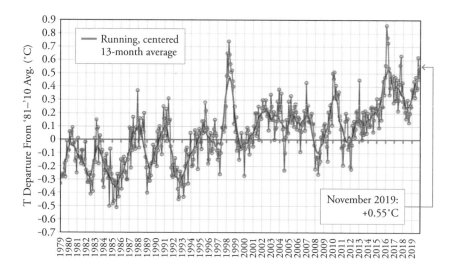

Source: Spencer 2019.

Figure 21. Satellite-based global temperature record, 1979–2019,
according to the Earth System Science Center at
the University of Alabama–Huntsville. Blue circles
and lines show monthly global-average temperature
anomalies for the lower troposphere, while the red line
is the running, centered 13-month average.

The Gap Between Observations and Models

In 2017, Christy calculated the five-year averaged values of annual tropical
midtropospheric temperature (TMT) variations for the years 1979–2016 as
depicted by 102 CMIP5 GCMs used by the IPCC to predict future global
temperatures. He then calculated the average of the models in thirty-two
institutional groups and plotted them in the figure that appears below as
Figure 22 (dotted lines). He then plotted the actual temperatures observed
for those same years, using green circles to show the average of four balloon
datasets; blue squares for three satellite datasets; and purple diamonds for
three reanalyses. The last observational point at 2015 is the average of 2013–16
only, while all other points are centered, five-year averages.

Christy's results reveal that all but one of the climate models produce

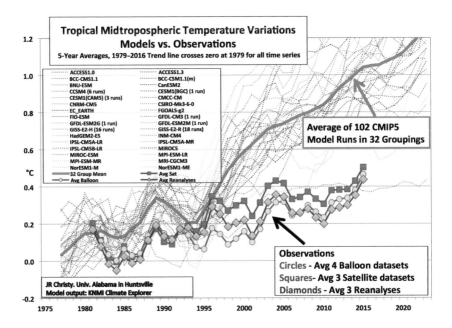

Source: Christy 2017.

Figure 22. Climate model forecasts versus observations, 1979–2016. See text or caption of Figure 11 in Chapter 7 for notes.

temperature estimates well above observed temperatures. Christy summarized his findings as follows:

When the "scientific method" is applied to the output from climate models of the IPCC AR5, specifically the bulk atmospheric temperature trends since 1979 (a key variable with a strong and obvious theoretical response to increasing GHGs in this period), I demonstrate that the consensus of the models fails the test to match the real-world observations by a significant margin. As such, the average of the models is considered to be untruthful in representing the recent decades of climate variation and change, and thus would be inappropriate for use in predicting future changes in the climate or for related policy decisions (Christy 2017).

Recall, too, from Chapter 4 the discussion of the "missing hotspot" that all climate models predict should exist in the tropical troposphere. When three colleagues and I compared observations to model predictions in this part of the atmosphere (called the Characteristic Emission Layer [CEL], the layer of the atmosphere between 450 and 750 hPa), we found virtually no overlap, as shown in Figure 4 in that chapter. We concluded our article in the *International Journal of Climatology* reporting these findings as follows:

> We have tested the proposition that greenhouse model simulations and observations can be reconciled. Our conclusion is that the present evidence, with the application of a robust statistical test, supports rejection of this proposition. (The use of tropical tropospheric temperature trends as a metric for this test is important, as this region represents the CEL and provides a clear signature of the trajectory of the climate system under enhanced greenhouse forcing.) On the whole, the evidence indicates that model trends in the troposphere are very likely inconsistent with observations, which indicates that, since 1979, there is no significant long-term amplification factor relative to the surface. If these results continue to be supported, then future projections of temperature change, as depicted in the present suite of climate models, are likely too high (Douglass et al. 2007).

McKitrick and Christy (2018) similarly tested the ability of GCMs to predict temperature change in the tropical troposphere, using model runs using the IPCC's Representative Concentration Pathway 4.5 (RCP4.5), which employs the best estimate of historical forcings through 2006 and anticipated forcings through 2100. According to the authors, "The mean restricted trend (without a break term) is 0.325 ± 0.132°C per decade in the models and 0.173 ± 0.056°C per decade in the observations. With a break term included they are 0.389 ± 0.173°C per decade (models) and 0.142 ± 0.115°C per decade (observed)." In other words, the models run hot by about 0.15°C per decade (0.325–0.173) and predict nearly twice as much warming in this area of the atmosphere as actually occurred (0.325 / 0.173) during the past sixty years. Similar results were reported by Monckton et al. in a peer-reviewed article appropriately titled "Why Models Run Hot: Results from an Irreducibly Simple Climate Model" (2015).

Of course, this is quite unexpected, since atmospheric concentrations of CO_2—the greenhouse gas that climate models presume to cause global warming—have been increasing rapidly in the twenty-first century. There have been many attempts to explain this discrepancy, ranging from a flat denial that such a gap exists (Karl et al. 2015) to attempts to account for the "missing incoming energy." For example, Kevin Trenberth has proposed that the missing energy is hiding in the deep ocean (Trenberth and Fasullo 2013). One possibility, of course, may be that the pause is simply a statistical fluctuation, like tossing a coin and getting fifteen heads in a row. Such an explanation cannot be dismissed out of hand, even though it has a very low probability, which becomes even smaller with each passing year of little or no warming. Obviously, climate alarmists like this possibility since it means no data can disprove their theory, but this only means their theory is not science, but only speculation.

The existence of the "pause" is creating a scientific challenge for climate skeptics and a real crisis for alarmists; it can no longer be ignored by any who consider themselves to be scientists nor, indeed, by responsible political leaders. Even if we cannot readily find the cause for the "pause," we can be absolutely sure that it was not predicted by any of the dozens of the IPCC's models. Therefore, logically, such nonvalidated GCMs cannot, and should not, be used to predict the future climate or as a basis for policy decisions.

Internal Causes

Most scientists are looking for a physical cause for the pause, an explanation of why the output of GCMs fails to match observations. When we look at possible causes, we should distinguish between internal and external ones that might offset the expected warming from CO_2.

Internal causes are negative feedbacks from either water vapor or clouds; they act to decrease the warming that should be attributed to increasing CO_2. The problem with internal effects is they, almost by definition, can never fully eliminate the primary cause. So even if they diminish the CO_2 effect somewhat, there should still be a remaining warming trend, though small.

It is quite important to obtain empirical evidence for a negative feedback. In the case of water vapor, one would look to see if the cold upper troposphere (UT) was dry or moist. If moist, as assumed implicitly in current IPCC

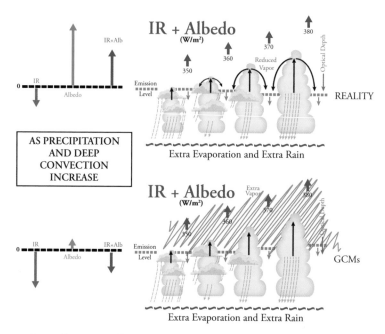

Source: Gray 2012, 9, figure 8.

Figure 23. How the continuous intensification of deep cumulus
convection would act to alter radiation flux to space.
The top diagram, labeled "Reality," emphasizes
the increasing extra mass flow return subsidence
associated with an ever-increasing depth and intensity
of cumulus convection. The bottom diagram, labeled
"GCMs," illustrates how GCMs interpret the increase
of deep convection as adding moisture to the upper
tropospheric levels and causing a decrease of radiation
to space. The bottom diagram is not realistic and is the
primary reason why the GCMs exaggerate the influence
of CO_2 on global warming.

GCMs, one gets a *positive* feedback; that is, an amplification of the CO_2-
caused warming. On the other hand, if the upper troposphere is dry, then
most emissions into space take place from water vapor in the warm boundary
layer in the lower troposphere. This leaves less energy available to be emitted
into space from the surface through the atmospheric "window," and therefore
produces a cooler surface. It would be a *negative* feedback.

The physical model I have in mind for this negative feedback is based on a proposal by William M. Gray, a great climatologist and expert on hurricanes at Colorado State University who passed away in 2016 (Gray 2012). (See Klotzbach et al. 2017 for a summary of his scholarly achievements.) Gray pictured cumulus clouds carrying moisture into the upper troposphere but occupying only a small area; as the water freezes out at higher altitudes, the dry air at that altitude gradually subsides back down to the surface, and (says Gray) that happens over a much greater area than the upwelling of moist air. (See Figure 23.) In principle, it should be possible to measure this difficult-to-explain effect fairly easily, using available satellite data. Gray's explanation matches the observed data better than the GCMs.

Negative feedback from increased cloudiness is easier to describe but more difficult to measure. The idea is simply that a slight increase in SST from the greenhouse effect of rising CO_2 also increases evaporation (according to the well-known "Clausius-Clapeyron" relation) and that this increased atmospheric moisture can also increase cloudiness. The net effect is a greater (reflecting) albedo, less sunlight reaching the surface, and therefore a *negative* feedback that reduces the original warming from increasing CO_2. (See Lindzen and Choi 2011.) Unfortunately, establishing the reality of this cloud feedback requires a measurement of global cloudiness with an accuracy of a small fraction of a percent, a very difficult problem.

External Causes

The principal *external* effects that might explain the existence of a global warming pause are aerosols and solar activity. Atmospheric aerosols, generally human-caused, can increase as well as decrease albedo and either cool or warm the planet, especially if they also increase cloudiness by providing condensation nuclei for water vapor. Contrails created in the wake of jet aircraft are one example.

The problem here is one of balancing: The net amount of cooling by aerosols, for example, has to be just right to offset the warming from CO_2 during the entire duration of the pause. It is difficult to picture why exactly this might be happening; the probabilities seem rather small. The burden is on the proponents to demonstrate various kinds of evidence in support of such an explanation.

Jet Contrails

Commercial and military air traffic inject billions of tons of water vapor as well as sulfur dioxide and nitrogen oxides into the upper atmosphere every year, leading to contrails and invisible cirrus clouds made up of ice crystals from the condensation of water vapor. The presence of such contrails is growing over time as air travel becomes increasingly common and jets fly at higher altitudes to avoid traffic. In many cities with busy airports, a permanent haze is now visible due to this phenomenon (Singer 1997a).

If jet contrails covered the entire sky, they would block the "atmospheric window" from 8 to 12 micrometers (microns) and would cause intense warming of the surface. This warming has nothing to do with CO_2. Back in 1988, when the possibility of a "nuclear winter" following a nuclear war was being debated, I concluded that atomic explosions on the ground would carry enough water vapor and dust into the stratosphere to cause (at least initially) intense warming. See Singer (1988).

More recently, Patrick Minnis and colleagues (2004) calculated that nearly all the surface warming observed over the United States between 1975 and 1994 (which they placed at 0.54°C) could be explained by aircraft-induced increases in cirrus cloud coverage over that period. Based on the three-day grounding of all commercial aircraft in the United States following the terrorist attacks on September 11, 2001, David J. Travis and colleagues (2012) found "an anomalous increase in the average diurnal temperature range (that is, the difference between the daytime maximum and night-time minimum temperatures) for the period September 11–14, 2001. Because persisting contrails can reduce the transfer of both incoming solar and outgoing infrared radiation, and so reduce the daily temperature range, we attribute at least a portion of this anomaly to the absence of contrails over this period." Later research by Jase Bernhardt and Andrew M. Carleton (2015) estimated that "contrail breakouts" reduced diurnal temperature range by between 2.8° and 3.3°C.

A phenomenon similar to jet contrails is ship tracks—bright streaks that form in layers of marine stratus clouds created by emissions from ocean-going vessels. These are persistent and highly reflective linear patches of low-level clouds that tend to cool the planet (Ferek et al. 1998). Mathias Schreier and colleagues (2006) concluded "modifications of clouds by international shipping can be an important contributor to climate on a local scale."

Solar Influences

Changes in the energy output of the Sun have long been known to influence Earth's climate. Various authors have linked observed climate parameters to the solar cycle, and temperature changes to the variation of solar-cycle length. (See Figure 24.) The Active Cavity Radiometer Irradiance Monitor (ACRIM) total solar irradiance (TSI) composite shows a small upward pattern from around 1980 to 2000, an increase not acknowledged by the IPCC or incorporated into the models on which it relies (Scafetta and Willson 2014). Even small changes in the absolute forcing of the Sun could result in values larger than the much smaller predicted changes in RF caused by human GHG emissions (Lindzen 2015).

Two TSI reconstructions for the period 1900–2000 by Scafetta and West (2006) suggest the Sun contributed 46 percent to 49 percent of the 1900–2000 warming of Earth, but with uncertainties of 20 percent to 30 percent in their sensitivity parameters. Close correlations exist between TSI proxy models and many twentieth-century climate records including temperature records of the Arctic and of China, the sunshine duration record of Japan, and the Equator-to-Pole (Arctic) temperature gradient record (Soon 2005, 2009; Ziskin and Shaviv 2012). The solar models used by the IPCC report less variability than other reconstructions published in the scientific literature (Soon, Connolly, and Connolly 2015), leading the IPCC to understate the importance of solar influences.

The conventional position is that the variation of the solar constant, only 0.1 percent during the solar cycle, is too small to have an effect. But solar variations also produce indirect climate changes through solar corpuscular radiation (solar "wind") sweeping past the Earth and the solar modulation of the

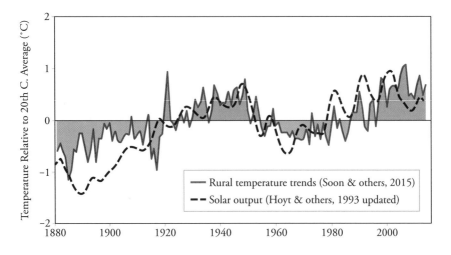

Source: Adapted from Soon, Connolly, and Connolly 2015, 442, figure 27.

Figure 24. Northern Hemisphere rural temperature trends versus solar output. Red and blue represent positive and negative temperature anomalies from twentieth-century average for a Northern Hemisphere temperature reconstruction using primarily rural surface stations (to control for urban heat island effect). Dashed line is solar output according to Hoyt and Schatten (1993) as updated by Scafetta and Willson (2014).

flux of cosmic rays that affect cloud formation, something I first wrote about many years ago (see Laster, Lenchek, and Singer 1962) and which has recently been experimentally confirmed (CERN 2016; Svensmark et al. 2017). This is an exciting frontier of climate science that is already leading some scientists to reduce climate sensitivity to CO_2.

Other Possible Causes

There is an important school of thought that does not rely on offsetting the forcing from increased CO_2; instead it assumes that there really exists an imbalance at the top of the atmosphere (TOA) and that global warming is taking place somewhere, but is not easily seen. Many assume that the "missing heat" is hiding in the deep ocean. It is difficult to see how such a mechanism can function without also raising surface temperatures, but an oscillation in ocean currents might produce such a result.

Still, if measurements could demonstrate a gradual increase in stored ocean heat, one would be forced to consider possible mechanisms. Its proponents might be asked why the storage increase started just when it did, when will it end, and how will the energy eventually be released, and with what manifestations?

There is yet another possibility worth considering: the missing energy might be used to melt ice rather than warm the ocean. Again, quantitative empirical evidence might support such a scenario. But how to explain the starting date of the pause, and how soon might it end?

* * *

The theoretical models used to predict future warming are not consistent with atmospheric observations; the present models cannot handle clouds and other important climate factors properly. While the heavily manipulated surface-based temperature record seems to approach the degree of warming predicted by these models, it is contaminated by local urban effects and covers only a small fraction of the globe. Satellite-based temperature readings are accurate and truly global, and they show a minor warming trend well below that predicted by the models.

The gap between computer models and observations remains an unsolved puzzle. The simplest description is that the climate sensitivity to CO_2 is close to zero, as demonstrated empirically. But why? Regardless of any unsettled science details, it seems sure that current climate models cannot represent what is actually happening in the atmosphere, and therefore one should not rely on their predictions. This discussion has important policy consequences since so many politicians are wedded to the idea that CO_2 needs to be controlled to avoid "dangerous" changes to the global climate.

Does CO2 Lead to Cooling?

UNTIL THE 1990s, it was generally understood that atmospheric carbon dioxide (CO_2) makes only a small contribution to global warming as a greenhouse gas. The Earth's hydrological cycle (the formation and destruction of clouds and precipitation systems around the Earth) is what controls the flow of atmospheric IR radiation through the troposphere. The thermodynamic properties of water in the atmosphere also assure us that under the range of temperatures and pressures found in the Earth's troposphere, water vapor's presence is self-limiting, meaning increasing the presence of other GHGs such as CO_2 will only enhance this hydrological cycle (meaning a slightly less humid atmosphere as more water vapor is used up in the hydrological cycle worldwide, which also increases cloudiness slightly and blocks solar shortwave energy). This cycle means any increase in the Earth's mean temperature from CO_2 or other GHGs aside from water vapor is negated and the planet's temperature remains stable.

But CO_2 is an interesting and complicated molecule. Its climate-forcing effect might actually decline to zero, albeit for only a number of years, because part of the CO_2 absorption and emission takes place in the stratosphere, where the temperature gradient is positive; that is, there is warming with increasing altitude, instead of cooling.

I believe that the gap between observations and model estimates will continue to grow in the future and demand a convincing empirical argument explaining why CO_2 no longer affects the climate, except perhaps at the slow level of its log-dependence. I propose, as a hypothesis, that CO_2, a greenhouse gas, can also cool the atmosphere. But until someone does the necessary work, by analyzing available satellite data, one should not put too much faith in this hypothesis.

The Theory

"Greenhouse gas" means only that CO_2 absorbs some IR radiation; it does not guarantee climate warming. In fact, the forcing of CO_2 depends on where it is in the atmosphere. Its actual behavior depends mostly on atmospheric structure, expressed by the atmospheric lapse rate (ALR). That is defined as change in atmospheric temperature with altitude, which has been measured by balloon-borne radiosondes. Figure 25 is an example of those measurements. The lapse rate is $-6.5°C/km$ low in the troposphere, but then goes to zero at the tropopause. In the stratosphere, the lapse rate is slightly positive.

The warming of the stratosphere is produced by absorption of energy from the Sun by stratospheric ozone. See the summary in Box 6.

Adding a tiny increment of CO_2 raises slightly the "effective" altitude for emitting outgoing long-wave radiation (OLR), the radiation going out to space from a CO_2 molecule. Because of the reversal in the atmospheric temperature structure, OLR is (1) of lower energy than normal if the effective altitude remains in the troposphere; and (2) a bit higher than normal if this effective altitude is in the stratosphere. In the second case, the stratospheric CO_2 emission "borrows" some energy from the surface emission, hence "cooling" the surface.

Box 6

Atmospheric Lapse Rate (ALR) by Level in Atmosphere

STRATOSPHERE	ALR is positive	Temperature increases with altitude
TROPOPAUSE	ALR is zero	Temperature is constant
TROPOSPHERE	ALR is negative	Temperature decreases with altitude

a. Logarithmic vertical scale

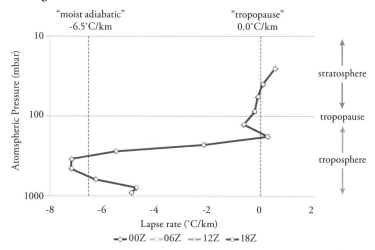

b. Linear vertical scale

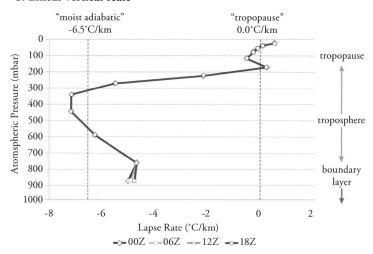

Source: Ronan Connolly and Michael Connolly, private comm., August, 2017.
See also Connolly and Connolly 2014.

Figure 25. Atmospheric Lapse Rate (ALR) by level. Lapse rate
determined from radiosonde measurements carried by
balloons. The data were taken in Ireland, variously at
midnight, 6 a.m., noon, and 6 p.m., and the four curves
in the figure are indistinguishable at all but extremely
low altitudes. Panel A shows pressure in logarithmic
scale, Panel B is linear scale.

This theoretical prediction is empirically confirmed by Arctic and Antarctic observations by Mark Flanner and colleagues (2018) and Antarctic observations by Holger Schmithüsen and colleagues (2015). The IR emission from the very cold surface has separated from the emission of the warmer stratosphere. This supposition can be checked by looking at spectral data from the atmospheric infrared sounder (AIRS) satellite. (AIRS is a satellite-borne IR spectrometer with ultrahigh resolution in wavelength.)

This cooling by CO_2 will reduce somewhat the normal greenhouse warming by CO_2, which will increase roughly as the logarithm of the CO_2 level in the atmosphere. We must also take into account the amplification by the (uncertain positive) water vapor feedback assumed by the GCMs used by the IPCC. But any CO_2 plus water vapor effects have historically (apparently) been overshadowed by climate oscillations such as the PDO and solar activity.

Assumptions

It is incumbent upon the proponent of a controversial hypothesis to find potential weak points and list crucial assumptions. To be sure, critics will soon enough find many more.

1. It seems safe to assume that CO_2 molecules, excited and de-excited by collisions with more abundant nitrogen and oxygen molecules, emit at the temperature of the surrounding atmosphere (van Wijngaarden and Happer 2018).

2. We may also assume that CO_2 is well mixed with altitude, as interhemispheric mixing is nearly perfect.

3. But can we assume that energy balance is nearly perfect, even on very short timescales? That is, will OLR always exactly equal absorbed solar shortwave radiation? I think the answer is *yes*. The timescales involved are too short to permit energy to exchange significantly between ice and ocean.

4. Most important, are CO_2 transitions strong enough to penetrate past the tropopause into the stratosphere? We can see evidence for this in the year-by-year increase of the amplitudes near the center of the 15-micron CO_2

absorption band. This increase comes about because the stratospheric ALR is positive. To verify and extend this observation, we may use data from the AIRS satellite instrument.

Once confirmed, the hypothesis can furnish additional explanation for the observed absence of CO2 warming in the past century and the puzzling observed warming "pause" of at least the past two decades.

A Typical Reaction

Physicists who have examined my counterintuitive hypothesis agree with the science—albeit somewhat reluctantly. Such is the power of groupthink that even experts, with some exception, find the idea that CO2 might cool the climate difficult to accept. One question that is raised is "where is the predicted cooling?" One can think of three possible answers:

1. First, the warming and cooling effects are very small; remember that the CO2 effect becomes logarithmic once the concentration exceeds roughly 60 ppm. The concentration is now 412 ppm, 0.04 percent, and growing.

2. Any cooling would be offset, at least partly, by molecular transitions that remain in the troposphere and cause climate warming in a conventional way.

3. Finally, there is *climate noise* that would hide any small warming or cooling. *Climate noise* is produced both naturally and by human sources. For example, changes in the weather may change global cloudiness and therefore incoming absorbed solar energy.

A GHG produces cooling of the climate when its molecular transitions are in a region of positive lapse rate. One example is CO2 in the stratosphere, where temperature increases with altitude. Another example is temperature over the winter poles. While the climate cooling is not obvious, it counters some conventional greenhouse warming. This at-least-partial cancellation might explain the puzzling absence of CO2-based greenhouse warming in the past century.

Much further work awaits!

11

Sea Level Rise

ONE OF THE most feared consequences of climate change is the possibility of a catastrophic sea level rise. Such a scenario is prominently featured in movies and documentary films such as Al Gore's *An Inconvenient Truth* and HBO's *Our Rising Oceans*. Since floods occur somewhere in the world nearly every week, producing heart-breaking news stories of lost homes and lives, they supply emotionally powerful images that climate change activists use to promote their views.

Sea Level Rise Is Not Accelerating

It is virtually impossible to predict (purely from theory) whether sea level will rise or fall as climate warms. On the one hand, melting glaciers and thermal expansion of ocean water will lead naturally to a rise in sea level (Wigley and Raper 1992). On the other hand, increased evaporation from the oceans and subsequent precipitation and accumulation of ice on Greenland and especially Antarctica would lower sea level (Oerlemans 1982). The only way to settle this issue is by examination of data.

Sea level has risen about 400 feet since the last glacial maximum of ~18,000 years ago, as shown in Figure 26. For the past several centuries, however, sea level is rising at the rate of only 1 to 2 mm per year. At that rate, sea level will be about 6 inches higher by 2100, a long way from Al Gore's 2006 estimate, in *An Inconvenient Truth*, of a twenty-foot rise.

Sea levels in many parts of the world have been rising and falling during past centuries for reasons that may have nothing to do with climate

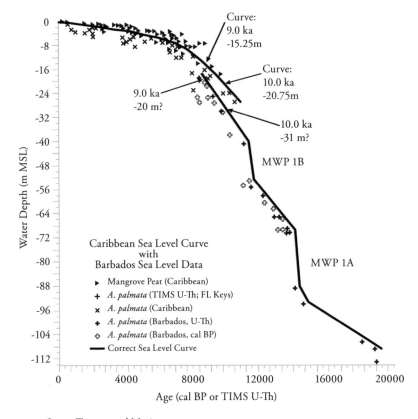

Source: Toscano and Macintyre 2003.

Figure 26. Sea level since last glacial maximum as deduced from coral and peat data. The graph is best understood by reading from the lower right (sea level 18,000 years before present) to the upper left (present sea level). The total rise for 18,000 years before present is about 120 meters. Note the rapid rate of rise as continental ice sheets melted and the more modest and nearly constant rate of rise in the past several millennia, irrespective of global temperature fluctuations.

change. Local relative sea level (LRSL) change is what matters most for coastal planning, and this is highly variable worldwide, depending upon the differing rates at which particular coasts are undergoing tectonic uplift or subsidence. The measure to focus on is *change in the rate* of sea level rise. In other words, has the long-term rate of sea level rise *accelerated* in the past century?

The best evidence on this question is tide gauge data from stations in tectonically stable areas with more than eighty years of uninterrupted recording. Such data show a steady linear sea level rise of about 18 cm per century and no acceleration in the past century (Parker and Ollier 2016, 2017). Figure 27 shows the record of sea level rise near three major coastal cities, none of which has experienced acceleration in the rate of sea level rise.

A Test of the Global Warming Theory

By studying a much shorter time interval, it is possible to sidestep most of the complications, like "isostatic adjustment" of the shoreline (as continents rise after the overlying ice has melted) and "subsidence" of the shoreline (as ground water and minerals are extracted). If the greenhouse theory is correct, rising levels of CO_2 should have caused a measurable acceleration in the sea level trend during the thirty-year period from 1915 to 1945, when a genuine, independently confirmed warming of approximately 0.5°C occurred. Yet the data plotted in Figure 28 clearly show that the rate of sea level rise was not affected by the warming during that period.

This conclusion is worth highlighting: Sea level rise is not affected by the use of fossil fuels. The evidence should allay fear that the release of additional CO_2 will increase sea level rise.

The Missing Water Is Turning into Ice

While there is reliable physical evidence showing sea levels rose at a *constant rate* during the entire past century, the cause of the trend (or rather, the absence of a change in the trend) is a puzzle. Physics demands that water expand as its temperature increases. But for the rate of rise to remain constant even as temperatures rise, as observed between 1915 and 1945, expansion of seawater must be offset by something else. What could that be? I conclude that it must be ice accumulation, through evaporation of ocean water and subsequent precipitation turning into ice. Evidence suggests that accumulation of ice on the Antarctic continent has been offsetting the steric effect for at least several centuries (Ligtenberg et al. 2013).

a. Honolulu, Hawaii, USA

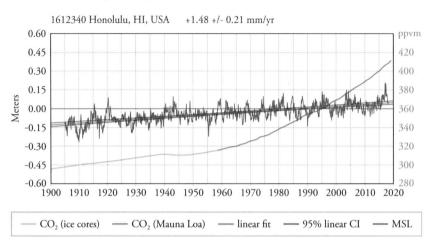

1612340 Honolulu, HI, USA   +1.48 +/- 0.21 mm/yr

b. Wismar, Germany

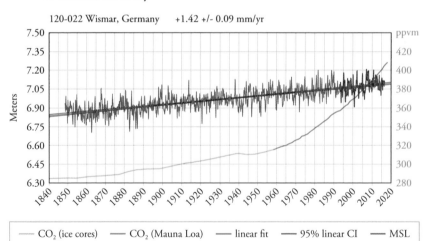

120-022 Wismar, Germany   +1.42 +/- 0.09 mm/yr

c. **Stockholm,** Sweden

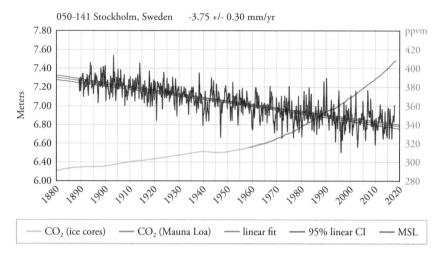

Source: Burton 2018.

Figure 27. Coastal measurement of sea level rise (blue) in three cities versus atmospheric CO2 concentration (green). Monthly mean sea level in meters (blue, left axis) without the regular seasonal fluctuations due to coastal ocean temperatures, salinities, winds, atmospheric pressures, and ocean currents. CO2 concentrations in ppmv (green, right axis). The long-term linear trend (red) and its 95 percent confidence interval (gray). The plotted values are relative to the most recent mean sea level data established by NOAA CO-OPS.

Melting of glaciers and ice sheets adds water to the ocean and causes sea levels to rise. (The melting of floating sea ice adds no water to the oceans, and hence does not affect the sea level.) After the rapid melting away of northern ice sheets following the end of the last Ice Age, the slow melting of Antarctic ice at the periphery of the continent may be the main cause of current sea level rise.

Melting is occurring presently at the Ross Ice Shelf of the West Antarctic Ice Sheet. Geologists have tracked Ross's slow disappearance, and glaciologist Robert Bindschadler predicts the ice shelf will melt completely within about 7,000 years, gradually raising the sea level as it goes (Bindschadler and

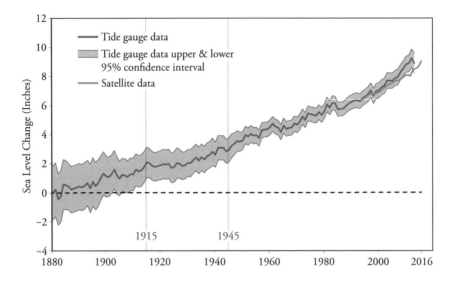

Source: US Global Change Research Program 2019.

Figure 28. Global average sea level change relative to 1880, 1880–2016, in inches. Notice the absence of any acceleration from 1915 to 1945, a period of known surface temperature warming.

Bentley 2002). Of course, a lot can happen in 7,000 years. The onset of a new glaciation could cause sea level to stop rising. It could even fall 400 feet, to the level at the last glaciation maximum 18,000 years ago.

All we know for sure is that currently, sea-level rise does not seem to depend on ocean temperature and certainly not on CO_2. We can expect the sea to continue rising at about the present rate for the foreseeable future. By 2100, the seas will have risen another 6 inches or so. The authors of the IPCC's assessments have recognized that actual measurements of sea level do not support earlier forecasts of acceleration and have reduced their estimates of projected sea level rise with each successive report, as shown in Figure 29. Much higher estimates of sea level rise created by James Hansen (Hansen and Sato 2012) and Stefan Rahmstorf (2007), also shown in Figure 29 as H and R, receive a lot of attention by the press but are far outside the maximum IPCC values. The ongoing rate of rise in recent centuries has been about 18 cm per

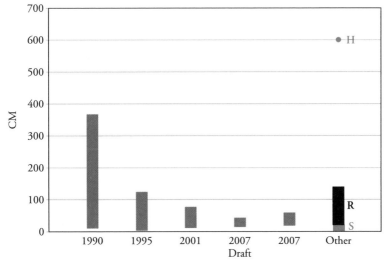

	IPCC 1990	IPCC 1995	IPCC 2001	2007 Draft	IPCC 2007	Hansen (H)	**Rahmst. (R)**	Singer (S)
Max	367	124	77	43	59	600	**140**	20
Min	10	3	11	14	18	600	**50**	18

Figure 29. Forecasts of sea level rise to year 2100 from IPCC reports of 1990, 1995, 2001, and 2007. Note the strong reduction in estimated maximum rise, presumably based on better data and understanding. Also shown are the published sea level rise values of Hansen and Sato (2012), Rahmstorf (2007), and Singer (1997b). Both Hansen and Rahmstorf are well outside the maximum IPCC values, which are converging on Singer's earlier estimate.

century; therefore, the incremental rate of rise (the rise attributable to global warming) for IPCC 2007 would be 0 to 41 cm, and about 0 to 2 cm for Singer.

* * *

A rise in sea level of about 6 inches in a century should not rank among the major problems facing mankind. The easy response is to simply stop repairing damaged buildings and infrastructure in low-lying coastal areas and allow a slow and almost costless retreat to higher ground. Or add a few inches to dikes and sea walls every few decades or so. Trying to lower sea levels by reducing our use of fossil fuels would be vastly more expensive and ultimately would not work.

12

Malthusians versus Cornucopians

SINCE THE PUBLICATION in 1798 of Thomas Robert Malthus's *An Essay on the Principle of Population*, some academics and others of note have expressed pessimism about the ability of mankind to feed and clothe its growing population. Their argument is simply that population, if left unchecked by war, hunger, or disease will always tend to outrun the growth of production. At first the availability of food was viewed as limiting the size of population. When this fear proved to be unfounded, exhaustion of natural resources and environmental degradation were added. The fact that such doomsday prophesies have proven false has not impaired the fervor of the Malthusians for coercive population control, nor shaken the faith of their followers.

Holding views largely opposite those of the Malthusians are the "Cornucopians," policy experts (mostly economists) who see the world much differently and, I would say, more clearly. Where the Malthusians look at population growth and see only more mouths to feed, Cornucopians see more brains to think and hands to work. Population growth leads to more innovation, which leads to higher productivity, and more hands lead to more economic growth. Economic growth in turn fuels the prosperity that instills many virtues, not least of which is a desire for a cleaner and healthier environment. Julian Simon (1995), Peter Huber (1999), and Indur Goklany (2007) are three prominent Cornucopians. (Simon passed away in 1998.)

The debate over climate change can be understood as just one more clash between the Malthusians and the Cornucopians. The same people who predicted gloom and doom in the 1960s and 1970s because of overpopulation, pesticides, and air and water pollution are now predicting a climate Armageddon, while those who (accurately) predicted growing global wealth and a cleaner environment are minimizing the possible risks associated with climate change. Understanding the rise of the environmental movement and some elementary economics sheds much needed light on the views and motivation of both sides.

Rise of the Environmental Movement

The almost unprecedented excitement and concern over climate change is due at least in part to the rise of the environmental movement. It is difficult to state exactly when global concerns first arose. Many would place the beginning in the late 1960s, after the appearance of Rachel Carson's *Silent Spring* (1962) and Paul Ehrlich's *The Population Bomb* (1968). An influential study at the time, now largely forgotten, was *Resources and Man*, produced by a committee led by Preston Cloud, a respected geologist and paleontologist, and published by the National Academy of Sciences in 1969.

The first celebration of Earth Day, in 1970, showed that the movement was already well underway. Certainly, passage of the National Environmental Policy Act and establishment of the US Environmental Protection Agency in 1970 in the United States advanced the cause. Other landmark events were the 1972 United Nations Conference on the Human Environment held in Stockholm, which launched the acid rain scare, and the publication of the Club of Rome's *Limits to Growth* (Meadows et al. 1972), fueling public concerns over the depletion of natural resources and particularly energy supplies. This apocalyptic work was wildly incorrect in its forecasts, but its use of computer models gave it the appearance of real science. The soberer and scientifically accurate book, *Inadvertent Climate Modification* (SMIC 1971), had relatively little popular impact. The 1971 controversy about supersonic transport and the fear of skin cancer from stratospheric ozone depletion aroused great excitement and further escalated global environmental concern. The Carter administration's gloomy *Global 2000: Report to the President* (Barney 1980), predicting that real food prices would double by the end of the 1990s, didn't hurt the cause either.

Since the 1980s, environmental activism has become a big business. Environmental groups fan fears of catastrophic climate change to raise billions of dollars a year. In 2012, 13,716 US environmental groups reported combined revenue of $7.4 billion and total assets of $20.6 billion (Nichols 2013). The Environmental Defense Fund (EDF) reported $112 million in revenues and $173 million in assets; Natural Resources Defense Council (NRDC) reported $97 million in revenue and $248.9 million in assets; and three tax-exempt Greenpeace organizations in the United States reported $39.2 million in revenue and $20.6 million in assets.

"Sustainable Development"

"Sustainable development" (SD) is the flag under which twenty-first century Malthusians march. The term was invented by Gro Harlem Bruntlandt, a Norwegian socialist politician and former prime minister. After her term there, she landed in Paris and, together with Club of Rome veteran Alexander King, began publicizing the concept. It masquerades as a call for clean air and "green" energy, and promises a pristine bucolic existence for us and our progeny—forever—if only we stop using fossil fuels, reject most of the advances of the Industrial Revolution, and severely reduce our consumption of virtually all things.

SD lives on because it is immensely useful to many groups who use the slogan to advance their own special agendas, whatever they may be. Some examples are:

- Restrictions on the use of fossil fuels
- Transfers of resources to less developed nations
- Striving for world government and UN sovereignty
- Promoting solar and wind power
- Advocating negative population growth

Among the worst policies being pushed with the help of SD is a scheme called Contraction and Convergence (C&C). The idea is that every human is entitled to emit the same amount of CO_2. This, of course, translates into every being on Earth using the same amount of energy—and, by inference, having the same income. In other words, C&C is basically a policy for a giant global income redistribution (Stott 2012).

Since the SD concept has been popularized, it has become a fashionable topic for research papers, especially in the social sciences. Trendy universities are establishing programs to teach SD—and even departments of SD and endowed academic chairs. Never underestimate the drive for expansion in the academic world.

For Earth Day 2011, the National Association of Scholars, composed mostly of conservative-leaning academics, sounded the alarm about the proliferation of SD courses in US colleges and universities. In a statement that critiques the campus sustainability movement, NAS president Peter Wood said, "Sustainability sounds like a call for recycling and clean drinking water. But its proponents are much more ambitious. For them, a sustainable society is one that replaces the market economy with top-down regulation. They present students a frightening story in which the earth is on the brink of disaster and immediate action is needed. This is a tactic aimed at silencing critics, shutting down debate, and mobilizing students who never get the opportunity to hear opposing views" (Wood 2011).

Resource Depletion

There is a grain of truth in the most preposterous claims of resource depletion: High-grade ores and easy-to-reach oil will become exhausted first—as they should under the commonsense principle that the lowest-cost resource is exploited before higher-cost resources are tackled. But the often-heard alarm that "reserves amount to only ten years' worth of annual use" doesn't mean that the resource will be exhausted in a decade—only that companies don't want to spend money looking for more reserves now; it's just unnecessary inventory on the shelf. These simple economic truths aren't always appreciated. In the late 1960s, at the height of the "oil crisis," the US National Academy of Sciences succumbed to depletion panic. Its report, *Resources and Man* (Cloud 1969), predicted the end of most metal resources by the year 1990—a bold statement but, again, quite wrong.

The surest signal of resource depletion is a rising price—and there is no sign of this as yet (even for crude oil, the most obviously depleting resource) (Clayton 2013). Clearly, improvements in extraction technology have been offsetting the poorer quality of resources coming on stream, whether ores or oil.

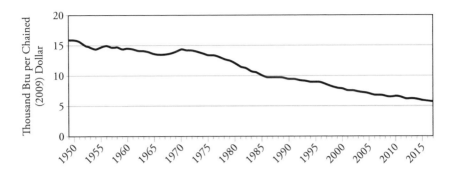

Source: EIA 2018, 16, figure 1.7.

Figure 30. US primary energy consumption per real dollar of gross domestic product (GDP), 1949–2017.

Conservation swings into action whenever prices begin to rise. We substitute paper for plastic (or plastic for paper) and silicon for copper. Recycling is routine for metals but is not possible for oil, coal, and gas, yet we can be confident that energy supplies will be plentiful even after oil prices itself out of the market. When fuel prices become too high, we substitute an equivalent but cheaper resource: hydropower, nuclear fission reactors fueled by uranium, and, in the future, breeder reactors fueled by plutonium, solar photovoltaic batteries, or perhaps nuclear fusion reactors and technologies not yet discovered.

Meanwhile, new products and manufacturing processes enable us to get more and more value out of every unit of natural resources we use, a gradual "dematerialization" of the global economy. New technologies have reduced the amount of energy required to produce a dollar of real gross domestic product (GDP) in the United States by two-thirds since 1949. (See Figure 30.)

Technological change will continue to reduce the energy intensity of the global economy in coming decades, partially offsetting the dramatic rise in demand for energy due to global population growth and rising prosperity. The environmental consequences of a growing global population would be far worse without innovation, as forests would need to be converted to cropland and emissions of all kinds, not just GHGs, would grow apace.

Environmental Protection

There is little dispute that environmental problems around the globe are caused by the waste products from growing populations and growing economies, and their increased consumption of resources, especially energy fuels. Problems of air, water, and solid pollution can justifiably be described as "pollution." But in countries where incomes are so low that staying alive is the main task, there is little left for pollution abatement and for the conservation of ecological resources, like forests or wildlife. We have a seeming paradox: as in the former USSR, insufficient economic output also leads to environmental degradation.

Even poor communities are willing to make sacrifices for some basic components of environmental protection, such as access to safe and clean drinking water and sanitary handling of human and animal wastes. As income rises, citizens raise their goals from mere survival to self-realization and spiritual goals. Once basic demands for food, clothing, and shelter are met, people demand cleaner air, cleaner streams, more outdoor recreation, and the protection of wild lands. With higher incomes, citizens place higher priorities on environmental objectives.

Ted Nordhaus and Michael Shellenberger (2007) wrote, "As Americans became increasingly wealthy, secure, and optimistic, they started to care more about problems such as air and water pollution and the protection of the wilderness and open space. This powerful correlation between increasing affluence and the emergence of quality-of-life and fulfillment values has been documented in developed and undeveloped countries around the world" (6). They continued, "Environmentalists have long misunderstood, downplayed, or ignored the conditions for their own existence. They have tended to view economic growth as the cause but not the solution to ecological crisis" (6).

Economists have documented what are called environmental Kuznets curves (EKCs) showing how various measures of environmental degradation rise with national per capita income until a certain tipping point and then begin to fall, often pictured as an inverted U shape (Panayotou 1993). Figure 31 shows a stylized rendition of the curve.

Gene M. Grossman and Alan Krueger (1995) conducted an extensive literature review of air quality over time and around the world and found

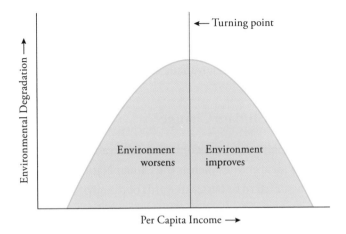

Source: Ho and Wang 2015, 42.

Figure 31. A typical environmental Kuznets curve. Environmental
   degradation rises with per capita income until a turning
   point is reached, and then decreases as additional
   income makes investments in environmental protection
   possible.

ambient air quality tended to deteriorate until average per capita income
reached about $6,000 to $8,000 per year (in 1985 dollars) and then began to
sharply improve. Later research confirmed similar relationships for a wide
range of countries and air quality, water quality, and other measures of environmental protection (Goklany 2007, 2012; Criado, Valente, and Stengos
2011; Bertinelli, Strobl, and Zou 2012).

   Still, the impact of population growth on environmental quality cannot
be dismissed out of hand; it requires detailed discussion, taking account of
the geographic scale—whether local, regional, or global—and the type of
pollutant, from sewage to toxic chemicals. Sewage depends strictly on the
number of people, but if their concentration is low, then natural processes can
absorb the pollution; if high, treatment plants must be built to avoid damage
to rivers and lakes. Chemical pollution and industrial pollution in general
require careful attention to manufacturing processes—a problem that can
be solved with the application of ingenuity and money.

In fact, ingenuity and money can solve most if not all environmental problems. If we can increase economic productivity and income, funds become available to abate pollution, whether through conservation, technological innovation, or mitigation of man-made changes to the environment.

Implications for Climate Change

The economic insights of Cornucopians are relevant to the climate change issue because political entities such as the United Nations or US government should not be assumed to be better stewards of the environment than private parties. The concerns of future generations are no better protected by politicians and voters today than they are by private asset managers and investors, and probably less so. The best responses to climate change are probably found in the private sector and not in the public sector.

Markets reward innovation, and innovation in turn benefits the environment. The best protection of the atmosphere rests in ensuring that technological innovations continue to increase humanity's ability to meet its material needs without further reducing the land available to wildlife or contaminating the planet's air and water.

The global nature of climate change and the fact that the planet's atmosphere is a *global commons* obscure the reality that the consequences of climate change are always experienced locally. Consequently, the information needed to anticipate changes and decide how best to respond is *local knowledge* and the most efficient responses will be *local solutions*. It often is forgotten that global estimates of temperature, sea level rise, and other measures of consequences are model-derived abstractions largely irrelevant to what occurs at specific locations around the world. Ross McKitrick (2001, 1) wrote,

> Anthropogenic additions to the atmosphere will (if they do anything) produce changes in the weather. But weather is a chaotic process, and we have limited expectation of being able to distinguish natural and anthropogenic changes at the local level, even *ex post*. Any damage function we define for the purposes of determining optimal mitigation policy must take for granted a future ability to accurately identify location-specific climate changes and attribute them to anthropogenic

causes. If we do not have this ability, climate policy cannot be based on cost-benefit analysis.

* * *

Most of the loudest voices in the climate change debate are those of Malthusians who previously and wrongly predicted mass starvation, environmental degradation, and resource scarcity. They have simplistically applied their pessimistic world view to another topic, climate change, failing to understand the complexity and scientific uncertainties that make predicting future climate conditions impossible.

Since the effect of reducing GHG emissions, according to the IPCC, will be only to delay the onset of global warming by a few months or years at best, global emission reduction programs are not an effective response to the real on-the-ground consequences of climate change even if one accepts the United Nations' scientific claims. The fact that the impacts of climate change are local explains why even managing the global commons that is the planet's atmosphere is best done by individuals and organizations throughout the world who are experiencing those impacts and not by international organizations based in New York, Paris, or The Hague.

13

Benefits of Modest Warming

IT SHOULD BE noted that little, if any, of the now more than $2 billion per year environmental research budget has been used to identify, document, or quantify possible benefits of adding carbon dioxide (CO_2) to the atmosphere, or of any of the other consequences of human activities. This bias has contributed greatly to public perception that these activities pose serious threats. However, that there are benefits from adding CO_2 to the atmosphere is undeniable.

How Much Warming Can We Expect?

Whether the benefits of global warming outweigh the costs depends crucially on *how much* and *how quickly* warming is likely to occur. All plants, animals, and human communities routinely adapt to small temperature changes and even to large changes if they occur slowly. Climate models are plainly not up to the task of predicting how much warming (or cooling) might lie ahead, but we can make a projection based on real-world observations. My best estimate of the *most* greenhouse warming likely to occur by the year 2100, based on assigning all of observed atmospheric warming to AGW, is 0.6°C. This estimate involves the following steps:

1. Satellite data show a temperature trend in the lower troposphere, after controlling for ENSO events, of about 0.08°C per decade.

2. Assume, conservatively, that all of this increase is due to increasing CO_2.

3. According to greenhouse theory, the surface trend should be about 20 percent less, or about 0.065°C/decade.

4. According to radiation theory, if CO_2 increases exponentially at the current rate, then the temperature trend will be linear.

5. So, by 2100, we should see an increase of $0.065 \times 9.5 = 0.6°C$, over the present value.

(Note that if CO_2 increases by 0.04 percent/year, then the value in 2100 will be 555 ppm; at 0.03 percent/year it will be 505 ppm versus preindustrial 280 ppm and the present 412 ppm.)

An increase in average global surface temperature of 0.6°C by 2100 would be too small to observe against background variability that, not counting the change of seasons, often exceeds that amount in a single year. This is much too small to justify forecasts of possible disasters (e.g., more floods, storms, and droughts). Smaller temperature increases yield correspondingly smaller climate impacts, and, in some cases, the possible harms turn into benefits. In fact, history and science and medicine all tell us that such a modest warming is likely to produce more benefits than harms both for humanity and for the natural environment.

Lessons from History

A large literature exists on the historical relationship between climate and human security (NIPCC 2019, chapter 7). Much of it shows humanity enjoyed periods of peace during warmer periods or periods of rising temperatures, while cooler periods or periods of falling temperatures have been accompanied by human suffering and often armed conflict. This research contradicts the narrative of the IPCC and its supporters, and for that reason it is seldom referenced in the IPCC assessment reports or by those who advocate for immediate action to address climate change.

China is a useful part of the world to study to learn the effects of climate change on human prosperity because it has been a well-populated, primarily agricultural country for millennia, and it has a relatively well-recorded history over this period.

A team of researchers led by Zhudeng Wei, a professor in the School of Geography at Beijing Normal University, investigated the long-term relationship between the climate and economy of China using a 2,130-year

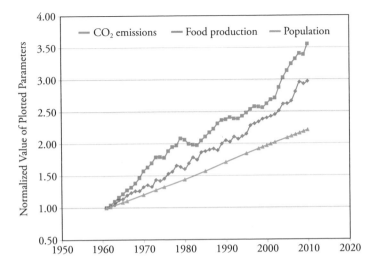

Source: Idso 2013, 24, figure 8.

Figure 32. Global population, CO2 emissions, and food
production from 1961 to 2010, normalized
to a value of unity at 1961. On the y axis, a
"normalized value" of 2 represents a value that
is twice the amount reported in 1961. Food
production data represent the total production
values of the forty-five crops that supplied 95
percent of the total world food production over
the period 1961–2011.

record they developed in previous research. They found that warm and wet
climate periods coincided with more prosperous and robust economic phases
(above-average mean economic level, higher ratio of economic prosperity, and
less intense variations), whereas opposite economic conditions ensued during
cold and dry periods, where the possibility of economic crisis was "greatly
increased" (Wei, Fang, and Su 2015).

Similarly, Haipeng Wang, a climatologist with the State Key Laboratory
of Cryospheric Science, a division of the Chinese Academy of Sciences, developed a 4,000-year proxy temperature reconstruction for Shanxi Province in
North China and compared it with published war and population records for
the province to explore the relationship between climate change and human

societal changes for this region. He found wars "occurred more frequently when temperature and precipitation decreased abruptly, and they also lasted for a relatively long time," and the most severe era of armed conflict occurred during the coldest period of the record; that is, the LIA (Wang et al. 2018, 388).

With respect to population, Wang and colleagues report "an increase [in population] often occurred during warm periods" (393), which provided relief from the harsh economic pressures brought about by poor crop harvests during colder periods, when yields were reduced by as much as 50 percent. Not surprisingly, reduced crop yields during cold eras would trigger higher food prices and famine, creating "large numbers of homeless refugees and outbreaks of plague," eventually resulting in "wars and social unrest which acted to reduce the population size" (392).

Research conducted in Europe and other parts of the world confirm what these Chinese researchers found: warm periods are beneficial for human populations while cold periods bring disaster in the form of crop failure and disease (Singer and Avery 2008; Tol and Wagner 2010). Weather should improve if greenhouse warming serves to reduce the equator-to-pole temperature gradient, as predicted. A warmer climate would mean fewer and less intense storms, benefiting farmers and reducing losses from natural disasters.

A Boon for Agriculture

The greatest impact of climate change, historically, has been on agriculture, with warm periods producing larger harvests and cold periods causing famines. A sustained warming, particularly with a reduced diurnal and seasonal temperature range, with warmer nights and milder winters, should benefit agriculture by extending the growing season. This will be aided by the fertilizing effect of CO_2 and the reduced need for water by plants exposed to higher levels of CO_2 (Idso 2013; NIPCC 2014).

The aerial fertilization effect of higher levels of atmospheric CO_2 has already increased food production. Contradicting forecasts of global famine and starvation by such popular figures as Paul Ehrlich and John Holdren (Ehrlich 1971; Ehrlich, Ehrlich, and Holdren 1977), the world's farmers increased their production of food at a faster rate than population growth, as shown in Figure 32.

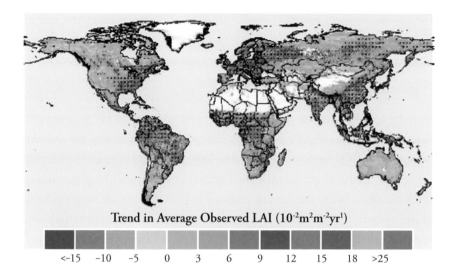

Trend in Average Observed LAI (10⁻²m²m⁻²yr¹)

$$<-15 \quad -10 \quad -5 \quad 0 \quad 3 \quad 6 \quad 9 \quad 12 \quad 15 \quad 18 \quad >25$$

Source: Zhu et al. 2016, figure S3.

Figure 33. Greening of the Earth, 1982–2009. The spatial distribution pattern of trend in growing season integrated Leaf Area Index (LAI) derived from global inventory modeling and mapping studies (GIMMS) LAI13g. LAI is the area of green leaf (one side) per unit of ground surface area. It is a dimensionless quantity (m^2m^{-2}) and shown here as $10^{-2}yr^{-1}$. Regions labeled by black dots indicate that those trends are statistically significant. Thirty-five percent of global vegetated area shows statistically significant increasing trends while 4 percent shows decreasing trends. Trend for entire period is 0.032 $m^2m^{-2}yr^{-1}$.

Growing global food production is resulting in less hunger and starvation worldwide. In 2015, the Food and Agriculture Organization of the United Nations (FAO) reported "since 1990–92, the number of undernourished people has declined by 216 million globally, a reduction of 21.4 percent" (FAO 2015, 8). In developing countries, the share of the population that is undernourished (having insufficient food to live an active and healthy life) fell from 23.3 percent twenty-five years earlier to 12.9 percent. A majority of the 129 countries monitored by FAO reduced undernourishment by half or more since 1996 (FAO 2015).

"Greening of the Earth"

Two decades ago, for the first edition of this book, I wrote of "growing evidence for the existence of a CO_2–fertilization effect, increasing the amount of biomass as CO_2 levels rise. Such a development would increase the fraction going into biomass and decrease the fraction of emitted CO_2 remaining in the atmosphere below the current 40 percent, thus slowing the rate of growth of atmospheric CO_2." I noted a "'greening' of the Earth at northern high latitudes has already been reported, as well as an earlier spring growing season."

Since then, evidence of a "greening of the Earth" has become overwhelming. The slight warming of the twentieth century and rising levels of CO_2 in the atmosphere have combined to cause a global and very beneficial greening Earth phenomenon, as shown in Figure 33. According to Zaichun Zhu and Shilong Piao, two professors of ecology at Peking University, in Beijing, China, global net primary production (NPP)—defined as the net carbon that is fixed (sequestered) by a given plant community or ecosystem—increased from 54.95 petagrams (Pg) C per year in 1961 to 66.75 Pg C per year in 2010, representing a linear increase of 21.5 percent over the period (Zhu et al. 2016).

Since CO_2 is the basic "food" of essentially all terrestrial plants, the more of it there is in the atmosphere, the bigger and better they grow. At locations across the planet, the increase in the atmosphere's CO_2 concentration has stimulated vegetative productivity (Cheng et al. 2017). This has beneficial effects for terrestrial animals as well, since it expands their habitat and enables humanity to grow the food it needs with fewer acres of farmland.

Positive Effects on Human Health

The mild warming of the past century was accompanied by a remarkable increase in the average lifespan of people living in nearly all parts of the world. According to the US Census Bureau (2016), "The world average age of death has increased by 35 years since 1970, with declines in death rates in all age groups, including those aged 60 and older. From 1970 to 2010, the average

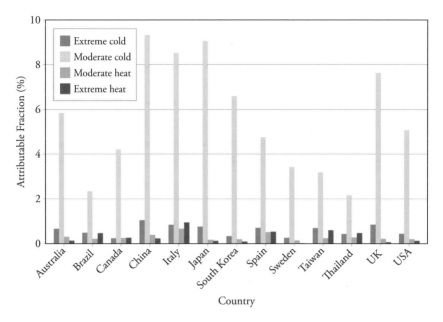

Source: Gasparrini et al. 2015, 369.

Figure 34. Deaths caused by cold versus heat, by country. Note that moderate cold contributes more by far to attributable deaths than moderate or even extreme heat.

age of death increased by 30 years in East Asia and 32 years in tropical Latin America, and in contrast, by less than 10 years in western, southern, and central Sub-Saharan Africa. . . . [A]ll regions have had increases in mean age at death, particularly East Asia and tropical Latin America" (31–33).

Medical science and observational research in Asia, Australia, Europe, and North America confirm that warming is associated with lower, not higher, temperature-related mortality rates (Gasparrini et al. 2015; White 2017). (See Figure 34.) Research shows warmer temperatures lead to decreases in premature deaths due to cardiovascular and respiratory disease and stroke occurrences (Nafstad, Skrondal, and Bjertness 2001; Gill et al. 2012; Song et al. 2018), while warmer temperatures have little if any influence on mosquito- or tick-borne diseases (Murdock, Sternberg, and Thomas 2016).

Of course, warmer temperatures alone were not responsible for the remarkable improvement in human health during the past century. The use of fossil fuels—the principal source of the CO_2 that presumably contributed to the warming trend during this period—played a major role in improving human health and longevity. Fossil fuels were responsible for the prosperity that lifted billions of people out of poverty, reducing the negative effects of poverty on human health (Moore and Simon 2000). They improved human well-being and safety by powering labor-saving and life-protecting technologies such as air conditioning, modern medicine, cars, trucks, and airplanes (Goklany 2007). Fossil fuels also increased the quantity and improved the reliability and safety of the food supply and make modern medicine possible (Moore and White 2016).

The current concerns about the spread of tropical diseases are certainly overblown when we consider that the most important vector in the spread of diseases is the human vector, aided by the growth and rapidity of global transportation. It would not be an exaggeration to state, as do Thomas Gale Moore (1995) and others, that climate change is good for you.

* * *

If the world is likely to see only 0.6°C warming in the coming century, then there is certainly no reason to tax or ration fossil fuels, subsidize alternative energies such as wind and solar power, or engage in the many other costly policies advocated by environmental activists and the policymakers who pay heed to them. People living in a modestly warmer world will be more prosperous, better fed, and healthier than people alive today, and they will enjoy a lusher and more bountiful natural world as well. What a pity it is that so few people know that this is the *real* story about climate change.

14

Mitigation, Sequestration, or Adaptation

WE TURN NOW to a key question in the climate change debate: What should we do about it? A reasonable answer in light of the science and economics presented in chapters before this one is "nothing at all." This answer is not popular with the many academics, leaders of nongovernmental organizations (NGOs), and government officials who have made careers out of advocating for immediate action.

Let us say, for the sake of argument, that the modest warming likely to occur in the century ahead merits action of some kind, if only to appease the general public that has been alarmed by all the hot talk coming from the alarmist side. There are three broad policy options: mitigation, sequestration, and adaptation. "Geoengineering," sometimes presented as a fourth option, can be used for sequestration or adaptation. A possible application of geoengineering to counter possible global *cooling* is presented in the next chapter.

Mitigation: Reducing Emissions

The response to climate change preferred by the IPCC and its many allies is to reduce human emissions of GHGs and, in particular, CO_2. The IPCC's Working Group II says GHG emissions must be cut by between 40 percent and 70 percent from 2010 levels by 2050 to prevent the ~2°C of warming (since preindustrial times) that would otherwise occur by that year (IPCC AR5 WGII 2014, 10, 12). There are three general ways to reduce emissions.

The first and most benign method is energy conservation and a more efficient use of energy through improved capital equipment or processes. In principle, conservation and energy efficiency save not only energy but also money; over time, saving money has been the main impetus behind such improvements. Problems arise only when efficiency increases are forced through arbitrary standards. A classic example is automobile fuel efficiency, which has increased up to a point because of a public demand for better gasoline mileage. The public balances that benefit with the loss of roominess, power, and safety that generally accompanies the downsizing and light-weighting of vehicles. The fuel economy standards adopted during the Obama administration were well beyond what the public wanted, resulting in increased purchases of pickup trucks and sports vehicles. It has been known for decades (Mayo and Mathis 1988) that those vehicles have comparatively poor fuel economy.

It is not generally recognized that conservation can be carried too far. Overconservation, which insists on replacing existing capital stock with more energy-efficient equipment, wastes energy—just like underconservation. It leads to the abandonment of equipment that is energy-imbedded and replaces it with equipment that requires energy to construct. As a general rule of thumb, one should not abandon equipment unless the energy savings from replacing it allow a payback in less than three to five years. If the payback period is too long, then energy is being wasted.

A different approach to reducing CO_2 emissions is to change to non-fossil-fuel sources of energy: hydroelectric, nuclear, solar and wind, and biofuels are the leading alternatives, each suited to satisfying some energy needs and not others. Figure 35 shows the contributions each fuel source makes to global energy supply. Solar and wind do not appear in Figure 35 because they supply too little power—just 6 exajoules (1.6 percent) in 2016—to appear in the figure.

Hydroelectric power sources are well established but require much energy to build, are very site-specific, and cannot be expanded indefinitely as demand grows. Furthermore, they have led to various ecological problems, particularly with fisheries. As a result, the Federal Energy Regulatory Commission (FERC) is now engaged in tearing down some small privately owned hydroelectric projects.

Nuclear energy is generally safer, less polluting, and sometimes cheaper than electric power generated any other way. Since the cost of nuclear electricity is mainly in the capital cost, rather than in the cost of the uranium fuel, economies in construction are particularly important. For example, in France,

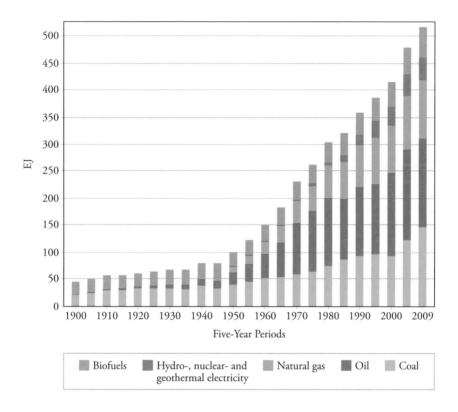

Source: Bithas and Kalimeris 2016, 8, figure 2.1.

Figure 35. The world's total primary energy supply for 1900–2009, measured in exajoules (EJ). Note the large and still growing contributions by fossil fuels and the absence of solar and wind, which are too small to appear in the graphic.

where plant design has been standardized and where the construction time has been compressed into five years or less, nuclear power is very economic. Of course, nuclear power has problems of its own. In the United States, litigation has often extended construction times to more than a decade, incurring huge interest costs. The public widely and unwisely fears nuclear power plants, believing they leak radiation and pose a threat of meltdown or explosion. In fact, they are the safest form of energy production currently available (Cohen 1984). In addition, the Nuclear Regulatory Commission has frequently required post facto construction changes based on insubstantial safety considerations, further raising the cost. Concerns of spent-fuel disposal and decommissioning of plants add very little to the cost but figure highly in the public debate.

Solar and wind energy are everyone's favorite; they seem to have few detractors. But their high capital cost makes them uneconomic except in specialized applications (Stacy and Taylor 2016). A special problem is their intermittency: energy must be stored to supply power at night, on cloudy days, and on days when the wind does not blow. Backup power generation must be available to ensure that electric grids are continuously powered. Despite massive subsidies, often justified by appealing to the alleged threat of climate change, it is unlikely solar and wind power will become a major supply source for electricity in the near future (Hughes 2012).

Biomass, in the form of wood and other materials, was the most common energy source before the arrival of fossil fuels and continues to be widely used today. But its potential as a replacement for fossil fuels is limited because it lacks the energy density required for electric power stations. Ethanol has lower power density than gasoline or diesel, corrodes pipelines and engine parts, and would require unacceptably large amounts of cropland to be devoted to producing energy rather than food to replace more than a small fraction of the oil used in transportation (Slade et al. 2011; Kiefer 2013).

The third method of limiting emissions, preferred by regulators, is to reduce the amount of fossil-fuel burning by rationing access to energy or taxing it. There are many schemes being discussed, including a hybrid scheme that includes both rationing and a form of taxation. There are difficulties with all of these schemes in deciding upon an equitable distribution of energy, in monitoring its use, and in enforcing limits on fuel burning. These problems are always present when rationing occurs and are only partially relieved when demand is limited by taxation.

A currently favored scheme, contained in the 2015 Paris Agreement, assigns an emission quota to each nation—called a Nationally Determined Contribution (NDC) by the UN, clearly a euphemism for rationing—but permits the buying and selling of unused emission permits, a kind of legalized black market. Ideally, this leads to the biggest reductions in emissions taking place where the cost of mitigation is lowest, resulting in a lower overall cost. The problems with such an approach are myriad. They begin with the initial allocation of national quotas. Should they be based on present energy consumption, or on the 1990 level, or on some hypothetical future level extrapolated from population growth? And should the per capita consumption of developing nations be set at some higher level than the present one—and if so, which level? The process is clearly political, and the resulting allocation of permits may do little to cut total global emissions (Babiker 2005).

Emission quotas and emission trading schemes are likely to create a permanent entitlement program that funnels money from industrialized nations needing emission permits to developing nations willing to sell them. It may even have the perverse effect of keeping developing nations from developing, if their governments decide that the transferred funds can be put to a "better" use, like building showy luxury projects or diverting it into foreign bank accounts. Even if the money is not squandered or misappropriated, it is likely to nurture a huge bureaucracy that could seriously throttle free enterprise and economic development in those nations (Kollmuss, Schneider, and Zhezherin 2015).

Emission trading is really a hidden energy tax for industrialized nations. Whoever buys the emission permits, whether utilities or the mining and drilling industries, must pass the cost along to someone—investors, workers, or consumers. Since economic growth is closely related to energy costs, the cost of reducing emissions by raising the price of fossil fuels includes the lost economic prosperity that otherwise would have occurred. When this factor is accounted for, reducing GHGs to 70 percent below 2010 levels by 2050 would lower world GDP in 2050 by 21 percent from baseline forecasts. World GDP would be about $231 trillion instead of the $292 trillion now forecast by the World Bank, a loss of $61 trillion (NIPCC 2019, SPM, 13). Worst of all, if emissions were to be limited to 70 percent below 2010 values—or even to values lower than that—atmospheric concentrations will still increase, albeit somewhat more slowly, given the long residence time of CO_2 in the atmosphere.

To summarize: Controlling GHG emissions by any method is extremely costly (Monckton 2016), distorts economic decisions, destroys jobs, is difficult to monitor, and practically impossible to enforce. It is likely to create huge international bureaucracies and police forces, damaging not only industrialized countries but certainly also energy exporters and most of the developing countries, since they depend on trade with the industrialized nations. And it would do little good unless emissions worldwide are cut drastically—80 percent and even 90 percent below baseline projections.

Sequestration: Storing CO2

The current excess of atmospheric CO_2 over its preindustrial value will eventually be absorbed by biota on land and in the ocean and stored, or sequestered, there. But even if a future warming is negligibly small and on the whole beneficial, there may still be political pressure to control the level of atmospheric CO_2. Speeding up the natural process of sequestering CO_2 could be a cost-effective alternative to reducing emissions.

Technologies used to remove CO_2 from the atmosphere are called negative emission technologies (NETs). NETs have been the subject of intensive study in recent years, spurred by realization that reducing emissions by the amounts called for by the IPCC and the Paris Agreement would be impossible or economically ruinous. Apparently some 2,900 studies on the idea were produced between 1991 and 2016, with almost 500 released in 2016 alone (Minx et al. 2017).

The best-studied scheme involves planting giant forest plantations that can extract CO_2 from the atmosphere (Shepherd 2009). The process is straightforward, in that one has to select fast-growing tree species and find locations where land costs and labor costs are reasonable. Unfortunately, quoted cost estimates vary widely and rise sharply as suitable land becomes scarcer. This is likely to happen because the areas involved are truly very large. If one uses as a rough guide one ton of carbon sequestered per hectare per year (Nordhaus 1991), absorbing current emissions would require an area of approximately 50 million square kilometers (4,500 × 4,500 miles). Although some attempts have been made by individual firms to plant forests that are said to offset their CO_2 emissions, forest-based sequestration of atmospheric CO_2 has not been pursued on a large scale (Boysen et al. 2016).

A lower-cost policy would be to use as much lumber as possible in all permanent structures and reseed existing forests. The carbon in the wood so used can be expected to remain in storage for fifty years or more, and its harvesting will spur fixation of additional carbon in the new trees grown to replace the old ones—without the need for additional acreage in forests.

A technique analogous to afforestation, but economically more attractive, is to speed up the natural absorption of CO_2 into the ocean. Currently, much of the world's oceans is a biological desert. Fertilizing parts of the oceans with nitrogen, phosphorus, and iron can cause algae blooms, which then die and carry the CO_2 they absorbed down to the ocean floor, where it is likely to remain forever. Recent research suggests such an effort could sequester around 15 percent (1.5 Pg C per year) of annual global CO_2 emissions (Harrison 2017). Costs and environmental risks could limit the scale of implementation. While it may never be necessary to reduce atmospheric CO_2, it will be comforting to know that we have the technical capability to do so via sequestration.

Adaptation: Turning Tragedy into Opportunity

A very different strategy, favored by many economists, is to adopt policies that allow people to adapt to any likely climate change. Adaptation to climate change, seasonal and interannual, has been the rule throughout human history. Populations have even adapted successfully to large permanent climate changes; for example, when Germanic tribes migrated from the frozen north to the Mediterranean (Singer and Avery 2008).

While adaptation to climate change may be problematic for some natural ecosystems, the ability to adapt is, paradoxically, highest for those economic sectors and human activities most sensitive to climate change. Because of their sensitivity to climate, such systems have always been heavily managed and have long histories of successful and rapid adoption of technological and management innovation. Besides energy conservation and the encouragement of non-fossil-fuel resources, actions meeting adaptation and development goals include increasing the hardiness, productivity, or efficiency of crops, livestock, forests, fisheries, and human settlements.

Adaptation is generally easier for technologically advanced societies and for societies with resources, which can afford adequate housing, heating, air conditioning, etc. Response strategies and impact assessment reports by the IPCC and other groups point out that developing countries are more vulnerable to climate change, not because climate change is expected to be greater in those nations—climate change will be least in the tropical zone—but because they lack financial and technical resources (e.g., IPCC AR5 WGIII 2014). Hence, it is imperative to expand the availability of these resources. This can be done through sustainable economic growth and technological change, which will reduce poverty and eventually population growth rates. These, in turn, require the establishment of the appropriate legal, economic, and institutional frameworks to encourage more economic growth and technological change.

The current excess of atmospheric CO_2 can become an important *resource* to be exploited for feeding a growing world population. For example, large-scale fertilization of areas of the Pacific and Southern Oceans for the purpose of stimulating the growth of phytoplankton would draw down atmospheric CO_2 without depressing the economies of industrialized nations or limiting the economic growth options of developing nations. With phytoplankton as the base of the oceanic food chain, any increase in that population can lead to the development of new commercial fisheries in areas currently devoid of fish. Carbon dioxide from burning fossil fuels thus becomes a natural resource for humanity rather than an imagined menace to global climate.

Economists describe how common resources can be degraded by overuse by "free riders," but also how they can be effectively managed by individuals and nongovernment organizations using their knowledge of local opportunities and costs, the kind of knowledge national and international organizations typically lack. These market-based solutions exhibit the sort of spontaneous order that Friedrich Hayek, the late Nobel Laureate economist, often wrote about (Hayek 1988), a coordination that is not dictated or controlled by a central planner. Elinor Ostrom, another Nobel Laureate, identified eight design principles shared by entities most successful at managing common-pool resources (Ostrom 2010).

The prosperity made possible by the use of fossil fuels has made environmental protection a social value in countries around the world (Hartwell and Coursey 2015). The value-creating power of private property rights, prices, profits and losses, and voluntary trade can turn climate change from a possible *tragedy* of the commons into an *opportunity* of the commons (Boettke 2009). Adaptation by free people, not government intervention, can balance the interests and needs of today with those of tomorrow.

* * *

Ignoring adaptation overestimates the negative impacts of climate change and overlooks opportunities to turn climate change, whether due to natural or man-made influences, into human benefits. Strategies aimed at control of CO_2 emissions from fossil-fuel combustion may compromise society's ability to cope with other global problems that require economic development. In contrast, successful adaptation to climate change requires specific actions—many of which will also help limit GHG emissions—that will stimulate sustainable economic growth and continued technological progress. Meeting these twin goals is critical to ensuring that limitation of GHGs, if it should become necessary, would cause the least disruption to society.

The most reasonable policy, then, is to adapt to climate change, as human activities normally adapt to seasonal and year-to-year variations in weather. One can then use the funds saved to strengthen the resilience of national economies, particularly in developing countries, against naturally occurring extreme climate events that cause damage (Goklany 1992, 1995).

15

Unfinished Business

CLIMATE SCIENCE IS not "settled"; it is both uncertain and incomplete. The available observations do not support the GCMs that predict a substantial global warming unless immediate action is taken to control GHG emissions. We need a targeted program of climate research to resolve a wide range of unanswered questions.

Unsettled Scientific Issues

There are still unsettled scientific issues in the climate debate. Here are seven:

1. The fate of anthropogenic carbon dioxide (CO_2) and its residence time in the atmosphere are uncertain: this includes its uptake into the ocean; the biological pump; the missing carbon sink. Predicting the future growth of atmospheric CO_2 requires more precise estimates of residence time and the amounts of fossil fuels available for energy production than are now available. Some researchers suggest the CO_2 level will rise to eight times its preindustrial level, while others doubt whether it will even double.

2. The temperature record of the last hundred years is of poor quality, with many discrepancies. Surface temperatures disagree with recent measurements from satellites and balloons. The "urban heat island" effect may skew the record. Ship observations are particularly questionable, with an unexplained abrupt cooling between 1900 and 1903.

3. GCMs cannot account for past observations: the unusual temperature

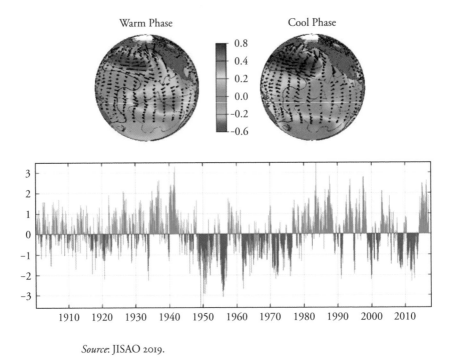

Warm Phase   Cool Phase

Source: JISAO 2019.

Figure 36. The Pacific Decadal Oscillation (PDO), 1900–
2017. Upper panel shows typical wintertime sea
surface temperature (colors), sea level pressure
(contours) and surface wind-stress (arrows).
Lower panel shows PDO index values (anomaly
patterns during warm and cool phases) from
January 1900 to January 2017.

rise from 1920 to 1940, the cooling to 1975, and the absence of warming in
the satellite and radiosonde records since 1979. Various reasons for these
discrepancies need to be explored before one places confidence in GCM
forecasts of warming.

4. GCMs vary by some 300 percent in their temperature forecasts, require
   arbitrary adjustments, and cannot handle crucial mesoscale and microscale
   cloud processes. Their forecasts of substantial warming depend on a
   modeled positive feedback from atmospheric water vapor (WV) that may
   not exist.

5. The effect of solar activity in climate change is still uncertain, though it

seems clear that GCMs understate its role. For example, the 1500-year cycle found in ice cores may be of solar origin, and cosmic rays may play a role in decadal cycles.

6. Similarly, the role of ocean currents—El Niño events, the PDO, North Atlantic Oscillation (NAO), and others—is still poorly understood. The satellite temperature data seem to reveal that El Niño events were responsible for virtually all the temperature increase since 1979. PDOs have "cold" and "warm" periods that last approximately twenty to thirty years, or full cycles of about sixty years (see Figure 36) while El Niño events last six to eighteen months (Trenberth and Hurrell 1994).

7. Sea level (SL) rise is a major feared result of future warming. But increased evaporation from oceans and more rapid accumulation of polar ice might lead to a lowering of sea level. This possibility is supported by an observed inverse correlation between SL rate of rise and tropical SST. I addressed this puzzle in an essay published by the *Wall Street Journal* on May 15, 2018. That article appears as Box 7 at the end of this chapter.

The Truth Needs to Be Told

In addition to unsettled scientific issues there are issues where the truth is known but it is not being shared with policymakers and the general public. Here are seven such issues:

1. Severe storms and hurricanes (except in the eastern North Pacific), as well as precipitation, appear to have diminished in the past fifty years. A calculated global warming trend, primarily at high latitudes, would reduce the latitudinal temperature gradient and therefore the driving force for storms and severe weather.

2. Global agriculture has benefited already from increased CO_2 and will continue to benefit from possible climate warming and increased precipitation, as increased nocturnal and winter warming leads to longer growing seasons. Increased CO_2 not only leads to more rapid growth of plants but also reduces their water requirements. Farmers can and will adjust to climate changes, whatever their cause.

3. The spread of disease vectors, like malaria-carrying mosquitoes, will be

unimportant in comparison to the growth in human vectors. Medical science and better insect control technology will overcome whatever diseases may be encouraged by warmer climates.

4. Historical evidence supports the idea that warmer climate intervals are beneficial for human activities, food production, and health. Cold periods have had the opposite effect.

5. Mitigation techniques are available that can slow down the rise of atmospheric GHGs and possible climate change: energy conservation and increased efficiency often make economic sense; hydro and nuclear power are available now, and mass construction in factories of safer, modular reactors would lower the cost of nuclear power by a factor of ten or more; regulatory preapproval could eliminate the enviro-caused delays that have been largely responsible for the current high cost of nuclear power.

6. Tree planting and ocean fertilization may be low-cost methods of sequestering atmospheric CO_2.

7. Attempting to mitigate climate change by imposing mandatory controls on energy use can create economic losses, harming especially poor people and poor nations, that exceed the possible benefits by orders of magnitude. Reducing atmospheric CO_2 levels by storing carbon in wood products or the deep oceans is possible but likely to be expensive and only partly effective. Adaptation—allowing or enabling people to change their behavior to accommodate changing temperatures and weather conditions—is generally superior to both mitigation and sequestration. Policy measures should be applied with great caution and only when justified by scientific data, lest they create more harm than good.

Preparing for Global Cooling

While governments and environmental activists focus on the possible threat of a trivial amount of warming, a bigger risk may be overlooked: global cooling. There are two kinds of ice ages: major glaciations and little ice ages. They are fundamentally different and therefore require different methods of mitigation.

Major Glaciations

According to the Serbian astronomer Milankovich, glaciation timing was controlled by astronomical parameters, such as oscillations with a 100,000-year period of the eccentricity of the Earth's elliptic orbit around the Sun; oscillations with a period of 41,000 years of the Earth's "obliquity" (inclination of the spin axis to the orbit plane, currently at around 23 degrees); and a precession of this spin axis, with a period of about 21,000 years.

The most recent major glaciation, ending only about 12,000 years ago, covered much of North America and Europe with miles-thick continental ice sheets and led to the disappearance of barely surviving bands of Neanderthals; they were displaced by the more adaptable *Homo sapiens*. While many consider the timing issue as settled, there are plenty of scientific puzzles still awaiting solutions: For example, how to explain the suddenness of deglaciation, transiting within only centuries from a glaciation maximum into a warm interglacial, like the present Holocene period.

Most expect the next glaciation to arrive rather soon, but calculations by Prof. Andre Berger of the Catholic University of Louvain, Belgium, suggest a delay of some 40,000 years—so there may be no great urgency (Berger 1988). Nevertheless, it would be useful and of great scientific interest to verify the existence of a hypothesized "trigger" that might be disabled by human action at low cost and negligible risk.

Little Ice Ages

Little ice ages and the Dansgaard-Oeschger Bond (DOB) cycles were discovered in Greenland ice cores by the Danish researcher Willi Dansgaard and by Swiss scientist Hans Oeschger, and further observed in ocean sediments by the late US geologist Gerard Bond. See *Unstoppable Global Warming: Every 1500 Years*, a book I coauthored with Dennis Avery in 2007, for a full account of this discovery.

We don't know what triggers an LIA, but suspect a strong correlation with a quiet Sun and prolonged absence of sunspots. Experts in this field—Willie Soon (Harvard-Smithsonian Center for Astrophysics), Harjit Ahluwalia

(University of New Mexico), Russian astronomer Habibullo Abdussamatov, and many others—believe the next LIA may be imminent. The most recent LIA lasted from AD 1400 to 1850—off and on. Its impact was rather severe. Climatology pioneer Hubert Lamb documents crop failures, starvation, and disease in Europe, together with ice fairs on the frozen Thames (Lamb 1985). During much of the American Revolution, New York Harbor was frozen over. And we recall paintings of George Washington crossing the Delaware River, impeded by ice floes.

How to Overcome a Little Ice Age

To avoid the huge human misery and economic damage, one would like to counteract the cooling phase of the DOB cycle—but how? The next LIA may be imminent, but there is no obvious trigger for solar influence and our understanding of solar physics is limited by the rather short history of observation of the Sun. While data on sunspots go back centuries, modern observations using spacecraft extend only for years.

One scheme to counter a cooling is to enhance greenhouse warming, perhaps by injecting substances into the atmosphere that absorb/reflect IR radiation. However, CO_2 is not the answer: CO_2 is limited in supply and is already saturated—hence additional CO_2 is not very effective. Synthetics, like sulfur hexafluoride (SF_6), are too long-lasting and may have risky side effects.

The answer may be water in the form of ice crystals. The scheme I have in mind is easily tested and is transitory—reversible and incurring little risk. A KC-135 or similar aerial-refueling aircraft carries ~100 tons of water, which is to be injected as mist just above the tropopause, at the bottom of the strato-sphere, near an atmospheric temperature minimum. At a surface density of water mist of 0.1 kg/m^2, the area covered would be ~1 km^2.

Like jet contrails discussed in Chapter 9, I expect some visible cirrus clouds to be created, which should disappear rapidly, leaving behind invisible cirrus ice crystals that are strong absorbers/emitters of IR radiation, covering also the atmospheric "window" region of 8 to 12 microns—thus creating a major GH effect and possibly even some detectable warming at the Earth's surface. Any satellite-borne IR instrument should be able to see this emitter patch and follow its spread and decay.

Of course, all this is based on theory and calculations, which (by the way) I published in 1988 in the peer-reviewed journal *Meteorology and Atmospheric Physics* during the debate over "nuclear winter" (Singer 1988). Obviously, all predictions must be validated by direct observations. Once the scheme is scientifically verified, operational planning for countering the cooling can take over.

As usual, there are many scientific questions that require answers—chiefly, understanding the physical mechanism that drives the DOB cycles and how to explain the size, shape, and duration of the abrupt quasi-periodic warmings. Currently, there is a hot dispute about the synchronicity of the cycles between the two polar regions, revolving about the limited accuracy of ice-layer dating in Antarctic cores. While the science is certainly interesting and important, there is no need to delay the crucial and urgent tests of geoengineering; they involve only minor costs and little risk to the atmospheric environment.

* * *

Probably the biggest lie in the climate debate is that "the science is settled." It is not. Major questions about temperature, causation, and impacts remain to be studied and better understood. Until this is done, the climate debate is far from over. Meanwhile, basic truths about the climate aren't being told: global temperature records are unreliable, the changes reported so breathlessly in newspaper headlines are trivial compared to natural variation, and many more. While governments focus on the remote threat of global warming, they may be overlooking the much bigger and more likely threat looming on the horizon: global cooling.

Box 7

I have been published hundreds of times in peer-reviewed scientific journals and even more often in newspapers and magazines. Almost alone among large-circulation newspapers, the Wall Street Journal *has been open to my work along with the work of other scientists who dissent from the Al Gore–Greenpeace alarmist line.*

"The Sea Is Rising, but Not Because of Climate Change"
By S. Fred Singer
Wall Street Journal
May 15, 2018

Of all known and imagined consequences of climate change, many people fear sea-level rise most. But efforts to determine what causes seas to rise are marred by poor data and disagreements about methodology. The noted oceanographer Walter Munk referred to sea-level rise as an "enigma"; it has also been called a riddle and a puzzle.

It is generally thought that sea-level rise accelerates mainly by thermal expansion of sea water, the so-called steric component. But by studying a very short time interval, it is possible to sidestep most of the complications, like "isostatic adjustment" of the shoreline (as continents rise after the overlying ice has melted) and "subsidence" of the shoreline (as ground water and minerals are extracted).

I chose to assess the sea-level trend from 1915–45, when a genuine, independently confirmed warming of approximately 0.5 degree Celsius occurred. I note particularly that sea-level rise is not affected by the warming; it continues at the same rate, 1.8 millimeters a year, according to a 1990 review by Andrew S. Trupin and John Wahr. I therefore conclude—contrary to the general wisdom—that the temperature of sea water has no direct effect on sea-level rise. That means neither does the atmospheric content of carbon dioxide.

This conclusion is worth highlighting: It shows that sea-level rise does not depend on the use of fossil fuels. The evidence should allay fear that the release of additional CO_2 will increase sea level rise.

But there is also good data showing sea levels are in fact rising at a constant rate. The trend has been measured by a network of tidal gauges, many of which have been collecting data for over a century.

The cause of the trend is a puzzle. Physics demands that water expand as its temperature increases. But to keep the rate of rise constant, as observed, expansion of sea water evidently must be offset by something else. What could that be? I conclude that it must be ice accumulation, through evaporation of ocean water, and subsequent precipitation turning into ice. Evidence suggests that accumulation of ice on the Antarctic continent has been offsetting the steric effect for at least several centuries.

It is difficult to explain why evaporation of seawater produces approximately 100 percent cancellation of expansion. My method of analysis considers two related physical phenomena: thermal expansion of water and evaporation of water molecules. But if evaporation offsets thermal expansion, the net effect is of course close to zero. What then is the real cause of sea-level rise of 1 to 2 millimeters a year?

Melting of glaciers and ice sheets adds water to the ocean and causes sea levels to rise. (Recall though that the melting of floating sea ice adds no water to the oceans, and hence does not affect the sea level.) After the rapid melting away of northern ice sheets, the slow melting of Antarctic ice at the periphery of the continent may be the main cause of current sea-level rise.

All this, because it is much warmer now than 12,000 years ago, at the end of the most recent glaciation. Yet there is little heat available in the Antarctic to support melting.

We can see melting happening right now at the Ross Ice Shelf of the West Antarctic Ice Sheet. Geologists have tracked Ross's slow disappearance, and glaciologist Robert Bindschadler predicts the ice shelf will melt completely within about 7,000 years, gradually raising the sea level as it goes.

Of course, a lot can happen in 7,000 years. The onset of a new glaciation could cause the sea level to stop rising. It could even fall 400 feet, to the level at the last glaciation maximum 18,000 years ago.

Currently, sea-level rise does not seem to depend on ocean temperature, and certainly not on CO_2. We can expect the sea to continue rising at about the present rate for the foreseeable future. By 2100 the seas will rise another 6 inches or so—a far cry from Al Gore's alarming numbers.

There is nothing we can do about rising sea levels in the meantime. We'd better build dikes and sea walls a little bit higher.

16

Conclusion

CLIMATE CHANGE IS a complex and difficult subject requiring the insights of many disciplines. It is easy to get lost in technical debates over the radiative properties of carbon dioxide (CO_2), for example, and overlook truths that have a more significant bearing on the debate. I suggest the reader pay attention to four essential truths revealed by the evidence presented previously:

- The warming from 1910 to 1945 was real: it is confirmed by thermometer records as well as proxy data, but it occurred *before* human greenhouse emissions could have caused it. The warming that may have occurred from 1978 to 1997 *is almost entirely fake*, an instrumental artifact found only in one heavily manipulated and unreliable database of surface observations.

- Since 2000, there has been little if any warming attributable to GHGs, a "pause" that is now approaching twenty years. (El Niño events in 1997–98 and 2016–2017 cannot be explained by CO_2 concentrations.) This means none of the extreme weather, floods, hurricanes, etc. that are so often attributed to "global warming" by the popular press and some prominent scientists could have been triggered by our GHG emissions. It is all fake news.

- GCMs fail to accurately replicate global temperatures since 1979 (when accurate satellite data became available); they "run hot," meaning they forecast more warming than has occurred in the past or will happen in the future. They are therefore unvalidated by observations, making them unsuited for use in policy making.

- The most reliable data on sea level show a steady linear rise of about 18 cm (about 7 inches) per century and no acceleration in the past century. The historical record shows the rate of sea level rise did not increase during the warming of 1910–45, demonstrating that the rate of sea level rise does not depend on air or sea-surface temperature. Therefore, predictions of increased coastal flooding or "disappearing islands" are not based on science, but instead are intended to frighten the public into supporting someone's political agenda.

If these four findings are true, it follows that there is no scientific reason to attempt to reduce the use of fossil fuels (the largest source of man-made CO_2 emissions is the combustion of coal, oil, and natural gas). Why, then, the focus by so many politicians, environmentalists, and scientists on the small CO_2 effect? The answer may be both political and scientific. The political aspect is obvious: politicians can control emissions of CO_2 by enacting taxes and imposing regulations on any activity that uses fossil fuels, which is just about everything. Politicians love control.

The scientific reasons for the narrow focus on CO_2 are more subtle. Scientific model builders are attracted to CO_2 because its climate effects, though tiny, can be calculated and allow construction of mathematical models. Many scientific careers have been launched and maintained by getting government grants to study this infinitely complex puzzle. Natural processes that have much larger effects on climate, such as solar activity and changing patterns of ocean currents, are essentially unpredictable by existing theory. Simply put, there's no money in studying natural causes of climate change.

There are still many things we don't understand about climate. We don't know why aerial concentrations of CO_2, a greenhouse gas, don't seem to affect the climate. The most likely explanation is that negative feedbacks and natural processes offset the gas's warming effect, or maybe it is offset by long-term solar cycles that are cooling the planet. But these feedbacks and processes are difficult to measure.

Similarly, we should expect the rate of sea level rise to accelerate due to the slow melting of glaciers and polar ice sheets that grew during the last LIA, on a time-scale of millennia, because it is warmer now than during the recent ice age glaciation more than 12,000 years ago. We also know water expands when heated. However, the sea level rise did not accelerate during 1910–45, a period when we are quite sure some (natural) warming took place. Something must be offsetting that expansion. I believe the offset comes from evaporation into the atmosphere and subsequent precipitation turning into ice over Antarctica. This explanation comports with evidence of increasing ice mass in Antarctica and raises questions about models claiming the opposite.

In an article published by the *Washington Times*, I said that "I have always been reluctant to make any predictions, 'especially about the future'" (Singer 2018). However, I went on to predict that the global warming "pause" of the last forty years would continue to the year 2100 and, likely, beyond. I predicted that the gap between GCMs and observed temperatures, so clearly documented in the work of John Christy and Roy Spencer, would also continue to grow. I also predicted that the global sea level would rise about 6 inches by 2100, too little to pose much of a risk to future generations.

The policy implications of all this were spelled out in Chapters 14 and 15, so I won't repeat them here. I will say that President Trump's decision to withdraw many Obama-era regulations, such as the Clean Power Plan, and to withdraw the United States from the Paris Agreement are entirely justified by the real science of climate change. Withdrawing from the FCCC and reversing the EPA's 2007 "endangerment finding," which labeled CO_2 a pollutant, ought to be on the agenda for the future.

Afterword

OVER THE PAST fifty years, climatology has undergone a radical and fundamental transformation. Before the 1970s, all the excitement and focus in atmospheric science was on meteorology, due to the then-recent advancements in numerical computations. Forecasting tomorrow's weather, studying hurricanes and tornadoes, understanding movements in the jet stream, learning about meteorological concepts through the newfangled satellites that we had recently put into space—this is what excited atmospheric scientists like me. Climatology, by contrast, was largely an actuarial science since the Earth's climate was simply defined by its "average weather." Climatologists were merely specialized statisticians who focused on the means, variances, and extremes of weather events. The Earth's climate, after all, was static and nonvariable—at least on the timescales of human lifetimes.

Then our view began to change. Scientific research led us to realize that the Earth's climate was indeed dynamic, variable, and ever-changing, and prominent climatologists began to note that global air temperatures were, in fact, *decreasing*. Were we headed toward another ice age? Books such as *The Cooling* by Lowell Ponte and an internal CIA study on the impact of climate change on intelligence problems, for example, highlighted that a changing climate could have dire consequences for all life on Earth, characterized by drought, famine, and political unrest. Global cooling had burst onto the scene.

But something happened on the way to climatological calamity—the Earth began to warm. The cooling that had become so disconcerting through the 1960s and early 1970s gave way to a concern over warming. Our newfound observations of the Earth from space helped shed light on the interaction between human activity and the Earth's climate. No longer were we concerned about the ills brought on by a globally cooler world; global warming became the new buzzword—or global heating, climate change, climate disruption, climate crisis, climate emergency, or whatever is the appropriate *mots justes* of the moment.

Global warming, however, was characterized by the same droughts, famines, and political unrest that were laid at the feet of its mirror-image doppelganger. The nouveau emphasis on environmentalism helped global warming capture the imaginations of activists everywhere, and global warming emerged at the forefront of their environmental concerns. Thus, the Earth's climate became a focal point for both science and public policy at the global scale.

The longevity of *Hot Talk, Cold Science* lies in that fact that very early on, Dr. Singer correctly grasped the concept that global warming is neither simply a scientific debate nor a political discussion; it is a horrid mixture of both. The first and second editions of the book were regularly criticized not for factual inaccuracies—indeed, it was frequently praised for its scientific accuracy and depth—but for being more about politics than science. Climate science is complex and difficult and, at times, fraudulent, but it also is muddied by the political environment in which we find ourselves. Many books were focused on the scientific questions surrounding global warming; many others have fixated on the politics. What Dr. Singer visualized in the First Edition of the book in 1997 was that global warming was a scientific question that would be adversely affected by the political scenario playing out on the international stage. As this Third Edition goes to press more than two decades later, the situation has not improved. In fact, the heated discussion of the science has only been exacerbated by the international geopolitical entities that seek to use global warming to further their agenda. *Hot Talk, Cold Science* was, and continues to be, about the nexus between science and politics; the two are inextricably intertwined.

From the famous Hansen hearing in 1988 and the rather subversive efforts of people like Undersecretary of State Timothy Wirth and IPCC Chairman Bert Bolin, global warming had been political since the beginning. The second IPCC Assessment Report (IPCC AR2), issued at about the time *Hot*

Talk, Cold Science was first published, was strongly one-sided and implicitly politically charged. A book like *Hot Talk, Cold Science* was sorely needed to elucidate the machinations of a political system that exploited global warming for political gains. Dr. Singer had the prescience to realize that global warming went far beyond the science that underlies it, and that global politics would ultimately dictate and even trump the science.

Why a Debate Exists

Skepticism is one of the hallmarks of science and the scientific method. The scientific method follows deductive reasoning where a hypothesis is evaluated by observations and experimental testing and ultimately refined, accepted, or rejected based on the findings. Rigorous skepticism is required for proper evaluation so that the hypothesis only passes through the scientific method if it can be demonstrated that the hypothesis is indeed valid, or at least, the hypothesis is not invalid.

Consider virtually any complex scientific question. Scientists will likely espouse a variety of disparate viewpoints because they tend to look at problems from different angles. The more complex the problem, the more and disparate the views are likely to be. Even the nature of gravity is hotly debated at the extremes—how it works at both the subatomic and galactic scales. Consequently, it is not surprising that a myriad of views on climate change will exist among honest scientists. Early analyses, such as the Charney Report in 1979 and, to some extent, the first IPCC Assessment Report (IPCC AR1), were more balanced by providing a variety of views and highlighting uncertainties.

Since then, however, alarmists have proclaimed "the science is settled" and subsequent IPCC reports (with some exceptions) and United States National Assessments have been more definitive and less tolerant of differing views. Scientific discussions in the popular press are likely one-sided and scientists who do not hold the alarmist viewpoint are treated with disdain, or worse. The reasons for this are quite varied, and it is difficult to pigeon-hole all alarmists. Many are infatuated with climate models. They embody the best science we think we know, but they often differ from observations, and our knowledge of climatology is still in its infancy, despite claims to the contrary (Chapter 9).

Many others, no doubt, become attached to the alarmist view because it is easier to go with the crowd. New PhDs are told by many of us in academia to remain quiet about their beliefs in climate change if they don't espouse the alarmist view. If young students want to gain tenure and have a future in academia, they cannot afford to express views that are not with the alarmist mainstream. We know from experience that many hiring decisions in academia are based on how much money one is likely to bring in and that tenure decisions usually are made on how successful one has been in bringing in that money. Comments by President Eisenhower in his Farewell Address (January 17, 1961) ring true even today:

> Akin to, and largely responsible for the sweeping changes in our industrial-military posture, has been the technological revolution during recent decades. In this revolution, research has become central, it also becomes more formalized, complex, and costly. A steadily increasing share is conducted for, by, or at the direction of, the Federal government. Today, the solitary inventor, tinkering in his shop, has been overshadowed by task forces of scientists in laboratories and testing fields. In the same fashion, the free university, historically the fountainhead of free ideas and scientific discovery, has experienced a revolution in the conduct of research. Partly because of the huge costs involved, a government contract becomes virtually a substitute for intellectual curiosity. For every old blackboard there are now hundreds of new electronic computers. The prospect of domination of the nation's scholars by Federal employment, project allocations, and the power of money is ever present—and is gravely to be regarded. Yet, in holding scientific research and discovery in respect, as we should, we must also be alert to the equal and opposite danger that public policy could itself become the captive of a scientific-technological elite.[1]

When Dr. Singer established the Science and Environmental Policy Project (SEPP) in 1990 to challenge flawed environmental policies enacted by the

---

1. D. Eisenhower, "Transcript of President Dwight D. Eisenhower's Farewell Address (1961)," Ourdocuments.gov (website), January 17, 1961, https://www.ourdocuments.gov/doc.php?flash=false&doc=90&page=transcript.

government using poor and biased science, he based its mission statement on the simple concept: omitting critical data violates the scientific method.

Activism, however, fuels much of the scientific-policy interaction highlighted in *Hot Talk, Cold Science*. Policy makers rely on extreme statements to garner an immediate response from scientists, and they usually have limited interest in or understanding of qualified statements or uncertainties. The late Dr. Stephen Schneider quipped that scientists often find themselves in a "double ethical bind"—bound by truth and its qualifiers but also by the need to obtain media coverage. These two issues easily may become at odds with each other:

> On the one hand, as scientists we are ethically bound to the scientific method, in effect promising to tell the truth, the whole truth, and nothing but—which means that we must include all the doubts, the caveats, the ifs, ands, and buts. On the other hand, we are not just scientists but human beings as well. And like most people we'd like to see the world a better place, which in this context translates into our working to reduce the risk of potentially disastrous climatic change. To do that we need to get some broad-based support, to capture the public's imagination. That, of course, entails getting loads of media coverage. So we have to offer up scary scenarios, make simplified, dramatic statements, and make little mention of any doubts we might have. This "double ethical bind" we frequently find ourselves in cannot be solved by any formula. Each of us has to decide what the right balance is between being effective and being honest. I hope that means being both.[2]

Many activist-scientists have failed at this ethical bind; however, extreme statements or blatantly false scenarios are never appropriate and are anathema to the scientific method. In discussing *An Inconvenient Truth*, Vice President Albert Gore commented:

> Nobody is interested in solutions if they don't think there's a problem. Given that starting point, I believe it is appropriate to have an over-

---

2. J. Schell, "Our Fragile Earth," *Discover*, October 1989, 47.

representation of factual presentations on how dangerous it is, as a predicate for opening up the audience to listen to what the solutions are, and how hopeful it is that we are going to solve this crisis. Over time that mix will change. As the country comes to more accept the reality of the crisis, there's going to be much more receptivity to a full-blown discussion of the solutions.[3]

Policy making is undermined and the scientific method is demeaned by such a blatant misrepresentation of science. Scientists today are encouraged to overstate the case for AGW either to "save the planet" or to possibly enhance their careers. That is why *Hot Talk, Cold Science* is so important.

Since Publication of the Second Edition

In the intervening two decades since the publication of the Second Edition, the politics and science of climate change have been dramatically transformed. According to the climate alarmists, the science is now "settled." They now recommend draconian policies to stave off their climate scares that have yet to materialize. Thirty-plus years of climate change hysteria have only risen to a crescendo and it shows no signs of abating.

We now await the release of the Sixth Assessment Report of the IPCC; each release has outdone the one that came before it. The United Nations, through its FCCC, has proposed several global agreements that focus on mitigation of, adaptation to, and penalties for GHG emissions. The twenty-fifth and latest annual meeting held in Madrid, Spain, in 2019 ended with only vague resolutions to reach the voluntary emissions targets set in Paris in 2016. The collectivist agenda that needs global warming policy to bring the countries of the world together suffered a major setback; however, they will not be deterred.

In the United States, President Donald J. Trump withdrew the United States from the Paris Agreement, although several states, municipalities, and universities have vowed that "we are still in!" Many on the political left are

---

3. D. Roberts, "Al Revere," Grist Magazine, May 9, 2006, http://www.grist.org/news/maindish/2006/05/09/roberts/index.html.

pushing a Green New Deal, which purports to address climate change and economic inequality through draconian economic and political legislation. Germany, England (pending what happens under Brexit), and other European Union (EU) member states also are grappling with climate change legislation and United Nations mandates that they have adopted. The EU has delivered its own version of the Green New Deal, but England has just exited the EU and Prime Minister Boris Johnson may take a more moderate view on climate change legislation. The climate change issue plays a major role in many democratic elections worldwide.

Scientific questions also have become much more complex. For example, wildfires in Australia and California recently have taken center stage and while they were initially blamed on global warming, they have been shown to result from bad forest management practices and several complex interactions among maintenance, weather, and land management factors. Sea level rise has continued to be a mainstay in the alarmist arsenal even though sea levels have risen steadily over the past hundred years, largely as a rebound from the LIA. After more than a decade (November 2005 through July 2017) without a landfall of a major hurricane (Category 3, 4, or 5) in the United States, the recent reappearance of landfalling major hurricanes is again being linked to AGW. A pause in global warming—the "hiatus"—that extended from the early 2000s to the mid-2010s did not deter the alarmists. Moreover, we have seen the emergence of the infamous "hockey stick," the impact of aerosols, the release of the "Climategate" emails, and the development of climate models as valid prognostications of the future. By contrast, the rise in polar bear populations has not caused their removal as the quintessential global warming icon.

Will It Ever End?

Will the political hype around global warming ever abate? In time, it will. But for the near and foreseeable future, global warming politics will continue to obscure the science behind it. Our fear is that another historical event where politics drove science will tell the tale of how the global warming science eventually becomes freed from its political trappings. The parallels are striking while the future appears quite bleak.

Beginning in the late 1920s under Joseph Stalin, Trofim Lysenko was the director of the Soviet Academy of Agricultural Sciences. At the time, Gregor Mendel's idea about genetics and the science that traits were inherited, some being dominant and others recessive, was taking shape. Lysenko and his disciples believed that environmental conditions experienced by the parents, and not the gene, were responsible for traits acquired by their offspring. Soviet Marxism was ready to accept Lysenko's views because Marxist propaganda held the view that scientific developments were more likely to arise from working-class people (like Lysenko) through experience. Marxism also was predisposed to the view that human traits were caused more by environmental determinism than by heredity; thus, socialism could fundamentally transform the populace which then would lead to ideas that would be inherited by future generations of their socialist comrades. Lysenko's ideas fit well with this goal.

History is littered with ideas that while they once seemed plausible, ultimately were consumed by the fires of the scientific method. Unfortunately, Lysenko and his followers were true science deniers—not only did they claim Mendel's genetic theory was wrong, to them it was anathema. Scientists that promoted the concepts espoused by Lysenkoism (a term coined by Vladimir Lenin) were lavished with governmental funding and given many awards to commend their efforts. By contrast, scientists who objected to Lysenko's views or found results that contradicted the antigenetic narrative were denounced as "bourgeois fascists" and had their reputations smeared as believers in a "bourgeoisie pseudoscience."

In August of 1948, the V. I. Lenin Academy of Agricultural Sciences proclaimed that Lysenkoism would be taught as the only correct theory. All Soviet scientists were required to accept Lysenkoism as scientific fact and to reject any research to the contrary. Scientists who studied genetics were deported, imprisoned, or executed. Lysenkoism became the scientific law of the Soviet Union and no scientist was permitted to hold or pursue any other viewpoint.

It was not until the transition of power from Nikita Khrushchev to Leonid Brezhnev in 1964 that the doctrine of Lysenkoism was finally repealed and scientists were free to pursue Mendelian genetics. But by then, Soviet agriculture was set back for a generation, and many Soviet citizens suffered as a result of this backward science. Other countries of the Eastern Bloc of nations and the People's Republic of China also were adversely affected by Lysenkoism and its implementation in the Soviet Union.

The parallels between the history of Lysenkoism and the current state of the politics surrounding global warming is striking. It took a whole generation to pass away before the fallacy of Lysenko's ideas and their adoption by a totalitarian regime could be dismantled. We fear that it may now take a whole generation of scientists to pass away before we can return to a state where climate change can be studied and evaluated in the light of true scientific inquiry and not from a politically correct perspective. We sincerely hope that we are wrong.

Summary

Hot Talk, Cold Science weaves the intricate tale of the interplay between the politics and the science that surrounds global warming. Although the politics and science of climate change now is far more complex than when the Second Edition of this book was written, both still affect governmental policies. An update to *Hot Talk, Cold Science* was sorely needed to include these recent developments, and no one was more qualified than Dr. Singer to write such a book. His myriad accomplishments serve to illustrate why he was one of the very few people who had both the requisite political experience and the scientific background. Over the years and despite all the invective poured upon him by the alarmists, Dr. Singer stood his ground for scientific integrity and adherence to the scientific method. For that, he is to be greatly commended.

DAVID R. LEGATES
Professor of Climatology, University of Delaware

ANTHONY R. LUPO
Professor of Atmospheric Science, University of Missouri

References

Ackerman, F., S. J. DeCanio, R. B. Howarth, and K. Sheeren. 2009. "Limitations of Integrated Assessment Models of Climate Change." *Climatic Change* 95: 297–315.

Aldrin, M., M. Holden, P. Guttorp, R. B. Skeie, G. Myhre, and T. K. Berntsen. 2012. "Bayesian Estimation of Climate Sensitivity Based on a Simple Climate Model Fitted to Observations of Hemispheric Temperatures and Global Ocean Heat Content." *Environmetric* 23 (3): 253–71.

Alley, R. B. 2000. "The Younger Dryas Cold Interval as Viewed from Central Greenland." *Quaternary Science Reviews* 19: 213–26.

Allison, I., N. L. Bindoff, R. A. Bindschadler, P. M. Cox, N. de Noblet, M. H. England, J. E. Francis, N. Gruber, A. M. Haywood, D. J. Karoly, G. Kaser, C. Le Quéré, T. M. Lenton, M. E. Mann, B. I. McNeil, A. J. Pitman, S. Rahmstorf, E. Rignot, H. J. Schellnhuber, S. H. Schneider, S. C. Sherwood, R. C. J. Somerville, K. Steffen, E. J. Steig, M. Visbeck, and A. J. Weaver. 2009. *The Copenhagen Diagnosis: Updating the World on the Latest Climate Science.* Sydney: The University of New South Wales Climate Change Research Centre (CCRC).

Armstrong, J. S., and K. C. Green. 2018. "Do Forecasters of Dangerous Manmade Global Warming Follow the Science?" Paper presented at the International Symposium on Forecasting, Boulder, Colorado, June 17–20.

Avery, S. K., P. D. Try, R. A. Anthes, and R. E. Hallgren. 1996. "An Open Letter to Ben Santer." *Bulletin of the American Meteorological Society* 77: 1961–6.

Babiker, M. H. 2005. "Climate Change Policy, Market Structure, and Carbon Leakage." *Journal of International Economics* 65: 421.

Balling Jr., R. C. 1992. *The Heated Debate: Greenhouse Predictions vs. Climate Reality.* San Francisco: Pacific Research Institute.

Barker, S. and Ridgwell, A. 2012. "Ocean Acidification." *Nature Education Knowledge* 3 (10): 21.

Barney, G. O. 1980. *The Global 2000 Report to the President.* Washington, DC: US Government Printing Office.

Bastin, J.-F., N. Berrahmouni, A. Grainger, D. Maniatis, D. Mollicone, R. Moore, C. Patriarca, N. Picard, B. Sparrow, E.-M. Abraham, K. Aloui, A. Atesoglu, F. Attore, C. Bassullu, A. Bey, M. Garzuglia, L. G. Garcia-Montero, N. Groot, G. Guerin, L. Laestadius, A. J. Lowe, B. Mamane, G. Marchi, P. Patterson, M. Rezende, S. Ricci, I. Salcedo, A. Sanchez-Paus Diaz, F. Stolle, V. Surappaeva, and R. Castro. 2017. "The Extent of Forest in Dryland Biomes." *Science* 356 (6338): 635–8.

Bell, Larry. 2011. "Climategate II: More Smoking Guns from the Global Warming Establishment." Forbes.com (website). November 29, 2011. https://www.forbes.com/sites/larrybell/2011/11/29/climategate-ii-more-smoking-guns-from-the-global-warming-establishment/#521c037e1323.

Berger, A. 1988. "Milankovitch Theory and Climate." *Reviews of Geophysics* 26 (4): 624–57.

Bernhardt, S., and A. M. Carleton. 2015. "The Impacts of Long-Lived Jet Contrail 'Outbreak' on Surface Station Diurnal Temperature Range." *International Journal of Climatology* 35 (15): 4529–38.

Bertinelli, L., E. Strobl, and B. Zou. 2012. "Sustainable Economic Development and the Environment: Theory and Evidence." *Energy Economics* 34 (4): 1105–14.

Bindschadler, R. A., and C. R. Bentley. 2002. "On Thin Ice?" *Scientific American* 287 (6): 98–105.

Bithas, K., and P. Kalimeris. 2016. *Revisiting the Energy-Development Link.* SpringerBriefs in Economics. https://doi.org/10.1007/978-3-319-20732-2_2.

Blair, T. A. 1942. *Climatology: General and Regional.* New York: Prentice-Hall.

Boettke, P. 2009. "Liberty Should Rejoice: Elinor Ostrom's Nobel Prize." The Freeman (website). November 18, 2009. https://fee.org/articles/why-those-who-value-liberty-should-rejoice-elinor-ostroms-nobel-prize/.

Bohm, R. 2012. "Changes of Regional Climate Variability in Central Europe During the Past 250 Years." *The European Physical Journal Plus.* https://doi.org/10.1140/epjp/i2012-12054-6.

Booker, C. 2009. *The Real Global Warming Disaster: Is the Obsession with "Climate Change" Turning Out to Be the Most Costly Scientific Blunder in History?* London: Continuum International Publishing Group.

Booker, C., and R. North. 1998. *Scared to Death: The Anatomy of a Very Dangerous Phenomenon.* London: Continuum International Publishing Group.

Boysen, L. R., W. Lucht, D. Gerten, and V. Heck. 2016. "Impacts Devalue the Potential of Large-Scale Terrestrial CO_2 Removal Through Biomass Plantations." *Environmental Research Letters* 11 095010.

Bryson, R. A., and W. M. Wendland. 1970. "Climatic Effects of Atmospheric Pollution." In *Global Effects of Environmental Pollution: A Symposium Organized by the American*

Association for the Advancement of Science held in Dallas, Texas, December 16–27, 1968, edited by S. F. Singer, 130–8. Berlin: Springer.

Burton, D. 2018. "Mean Sea Level at Honolulu, HI, USA (NOAA 1612340, 760-031, PSMSL 155), Mean Sea Level at Wismar, Germany (NOAA 120-022, PSMSL 8), and Mean Sea Level at Stockholm, Sweden (NOAA 050-141, PSMSL 78). Sea Level Info (website). Accessed December 11, 2018. http://www.sealevel.info/.

Business Today [India]. 2019. "One Billion Tonne Coal Production Would Be Achieved Soon: Coal Minister." October 18, 2019.

Caldeira, K., and J. F. Kasting. 1992. "Susceptibility of the Early Earth to Irreversible Glaciation Caused by Carbon Dioxide Clouds." *Nature* 359: 226–8.

Callendar, G. S. 1938. "The Artificial Production of Carbon Dioxide and Its Influence on Temperature." *Quarterly Journal of the Royal Meteorological Society* 64: 223.

Carilli, J., S. D. Donner, and A. C. Hartmann. 2012. "Historical Temperature Variability Affects Coral Response to Heat Stress." *PLOS ONE* 7: e34418. https://doi.org/10.1371/journal.pone.0034418.

Carson, R. 1962. *Silent Spring*. New York: Houghton Mifflin.

CERN. 2016. "CLOUD Experiment Sharpens Climate Predictions." *CERN Courier*. November 11, 2016.

Charney, J. G., A. Arakawa, D. J. Baker, B. Bolin, R. E. Dickinson, R. M. Goody, C. E. Leith, H. M. Stommel, and C. I. Wunsch. 1979. *Carbon Dioxide and Climate: A Scientific Assessment*. National Research Council. Washington, DC: The National Academies Press.

Cheng, L., L. Zhang, Y.-P. Wang, J. G. Canadell, F. H. S. Chiew, J. Beringer, L. Li, D. G. Miralles, S. Piao, and Y. Zhang. 2017. "Recent Increases in Terrestrial Carbon Uptake at Little Cost to the Water Cycle." *Nature Communications* 8: 110.

Christy, J. R. 2011. Testimony to the US House Science, Space, and Technology Committee. March 31, 2011.

———. 2012. Testimony to the US Senate Environment and Public Works Committee. August 1, 2012.

———. 2017. Testimony before the US House Committee on Science, Space, and Technology. March 29, 2017.

Christy, J. R., and R. T. McNider. 2017. "Satellite Bulk Tropospheric Temperatures as a Metric for Climate Sensitivity." *Asia-Pacific Journal of Atmospheric Sciences* 53 (4): 511-8.

Christy, J. R., R. W. Spencer, W. D. Braswell, and R. Junod. 2018. "Examination of Space-Based Bulk Atmospheric Temperatures Used in Climate Research." *International Journal of Remote Sensing* 39 (11): 3580–607.

Clayton, B. 2013. "Bad News for Pessimists Everywhere." Energy, Security, and Climate (blog). Council on Foreign Relations. March 22, 2013. https://www.cfr.org/blog/bad-news-pessimists-everywhere-malthus-was-wrong.

Cloud, P. 1969. *Resources and Man*. National Academy of Sciences. Washington, DC: US Printing Office.

Cohen, B. L. 1984. "Nuclear Power Economics and Prospects." In *Free Market Energy: The Way to Benefit the Consumer*, edited by S. F. Singer, 218–51. New York: Universe Books.

Connolly, M. and R. Connolly. 2014. "The Physics of the Earth's Atmosphere I. Phase Change Associated with Tropopause." *Open Peer Review Journal* 19 (Atm. Sci.), ver. 0.1 (non-peer-reviewed draft). http://oprj.net/articles/atmospheric-science/19.

Cook, E. R., R. Seager, R. R. Heim Jr., R. S. Vose, C. Herweijer, and C. Woodhouse. 2010. "Mega-droughts in North America: Placing IPCC Projections of Hydroclimatic Change in a Long-term Palaeoclimate Context." *Journal of Quaternary Science* 25: 48–61.

Costella, J., ed. and annotator. 2010. *The Climategate Emails*. Victoria, Australia: The Lavoisier Group.

Criado, C. O., S. Valente, and T. Stengos. 2011. "Growth and Pollution Convergence: Theory and Evidence." *Journal of Environmental Economics and Management* 62 (2): 199–214.

Dahl-Jensen, D., K. Mosegaard, N. Gundestrup, G. D. Clow, S. J. Johnsen, A. W. Hansen, and N. Balling. 1998. "Past Temperatures Directly from the Greenland Ice Sheet." *Science* 282 (5387): 268–71.

Darwall, R. 2014. *The Age of Global Warming: A History*, reprint. Northampton, MA: Interlink Publishing Group.

Douglass, D. H., and J. R. Christy. 2009. "A Climatology Conspiracy?" *American Thinker* (website). December 20, 2009. https://www.americanthinker.com/articles/2009/12/a_climatology_conspiracy.html.

Douglas, D. H., J. R. Christy, B. D. Pearson, and S. F. Singer. 2008. "A Comparison of Tropical Temperature Trends with Model Predictions." *International Journal of Climatology* 28 (13): 1693–1701.

Ehrlich, P. 1971. *The Population Bomb*, rev. and expanded ed. New York: Ballantine Books.

Ehrlich, P., A. Ehrlich, and J. Holdren. 1977. *Ecoscience: Population, Resources, Environment*, 3rd ed. New York: W. H. Freeman.

EIA (US Energy Information Administration). 2018. *Monthly Energy Review*. July, 2018.

Eisenhower, D. 1961. "Transcript of President Dwight D. Eisenhower's Farewell Address (1961)." Ourdocuments.gov (website). January 17, 1961. https://www.ourdocuments.gov/doc.php?flash=false&doc=90&page=transcript.

Ekblom, J. 2019. "EU Fails to Set Tougher Climate Targets Before December U.N. Conference." Reuters (website). October 18, 2019. https://www.reuters.com/article/us-eu-summit-climate-idUSKBN1WX1WC.

Engel, Z., K. Láska, D. Nývlt, and Z. Stachon. 2018. "Surface Mass Balance of Small Glaciers on James Ross Island, North-Eastern Antarctic Peninsula, During 2009–2015." *Journal of Glaciology* 64: 349–61.

EPA (US Environmental Protection Agency). 2016. "Climate Change Indicators: High and Low Temperatures." US Environmental Protection Agency (website). Accessed December 6, 2018. https://www.epa.gov/climate-indicators/climate-change-indicators-high-and-low-temperatures.

EPRF (Energy Policy Research Foundation, Inc.). 2011. *Implementation Issues for the Renewable Fuel Standard—Part I. Rising Corn Costs Limit Ethanol's Growth in the Gasoline Pool*. April 28, 2011.

Essex, C., R. McKitrick, and B. Andresen. 2007. "Does a Global Temperature Exist?" *Journal of Non-Equilibrium Thermodynamics* 32 (1): 1–27.

Fall, S., A. Watts, J. Nielsen-Gammon, E. Jones, D. Nivogi, J. R. Christy, and R. A. Pielke Sr. 2011. "Analysis of the Impacts of Station Exposure on the U.S. Historical Climatology Network Temperatures and Temperature Trends." *Journal of Geophysical Research* 116: https://doi.org/10.1029/2010JD015146.

FAO (Food and Agriculture Organization of the United Nations). 2015. *The State of Food Insecurity in the World 2015*. Rome: Food and Agriculture Organization of the United Nations.

FCCC (United Nations Framework Convention on Climate Change). 1992. New York: United Nations. http://unfccc.int/resource/docs/convkp/conveng.pdf.

Ferek, R. J., D. A. Hegg, P. V. Hobbs, P. Durkee, and K. Nielsen. 1998. "Measurements of Ship-Induced Tracks in Clouds Off the Washington Coast." *Journal of Geophysical Research* 103: 23,199.

Ferguson, R. 2007. *A Fundamental Scientific Error in "Global Warming" Book for Children*. Washington, DC: Science & Public Policy Institute.

Flanner, M. G., X. Huang, X. Chen, and G. Krinner. 2018. "Climate Response to Negative Greenhouse Gas Radiative Forcing in Polar Winter." *Geophysical Research Letters* 45 (4): 1997–2004.

Fountain, A. G., B. Glenn, and T. A. Scambos. 2017. "The Changing Extent of the Glaciers Along the Western Ross Sea, Antarctica." *Geology* 45: 927–30.

Gasparrini, A., Y. Guo, M. Hashizume, E. Lavigne, A. Zanobetti, J. Schwartz, A. Tobias, S. Tong, J. Rocklöv, B. Forsberg, M. Leone, M. De Sario, M.L. Bell, Y.-L. Leon Guo, C. Wu, H. Kan, S.-M. Yi, M. de Sousa Zanotti Stagliorio Coelho, P. Hilario Nascimento Saldiva, Y. Honda, H. Kim, and B. Armstrong. 2015. "Mortality Risk

Attributable to High and Low Ambient Temperature: A Multicountry Observational Study." *The Lancet* 386: 369.

Gill, R. S., H. L. Hambridge, E. B. Schneider, T. Hanff, R. J. Tamargo, and P. Nyquist. 2012. "Falling Temperature and Colder Weather Are Associated with an Increased Risk of Aneurysmal Subarachnoid Hemorrhage." *World Neurosurgery* 79: 136–42.

Goklany, I. M. 1992. "Adaptation and Climate Change." Paper presented at the annual meeting of the American Association for the Advancement of Science, Chicago, February 6–11, 1992.

———. 1995. "Strategies to Enhance Adaptability: Technological Change, Sustainable Growth and Free Trade." *Climatic Change* 30: 427–49.

———. 2007. *The Improving State of the World: Why We're Living Longer, Healthier, More Comfortable Lives on a Cleaner Planet*. Washington, DC: Cato Institute.

———. 2012. "Humanity Unbound: How Fossil Fuels Saved Humanity from Nature and Nature from Humanity." *Cato Policy Analysis* #715. Washington, DC: Cato Institute.

Goldstein, L. 2009. "Botch After Botch After Botch." *Toronto Sun*. November 29, 2009.

Gouretski, V., J. Kennedy, T. Boyer, and A. Köhl. 2012. "Consistent Near-Surface Ocean Warming Since 1900 in Two Largely Independent Observing Networks." *Geophysical Research Letters* 39 (19): L19606.

Gray, W. M. 2012. "The Physical Flaws of the Global Warming Theory and Deep Ocean Circulation Changes as the Primary Climate Driver." Powerpoint presentation at the Seventh International Conference on Climate Change, Chicago, May 21–23, 2012.

Graybill, D. A., and S. B. Idso. 1993. "Detecting the Aerial Fertilization Effect of Atmospheric CO_2 Enrichment in Tree Ring Chronologies." *Global Biogeochemical Cycles* 7: 81–95.

Grossman, G., and A. Krueger. 1995. "Economic Growth and the Environment." *Quarterly Journal of Economics* 110 (2): 353–77.

Hahn, R. W., and P. C. Tetlock. 2008. "Has Economic Analysis Improved Regulatory Decisions?" *Journal of Economic Perspectives* 22 (1): 67–84.

Hansen, J., and M. Sato. 2012. "Paleoclimate Implications for Human-Made Climate Change." In *Climate Change: Inferences from Paleoclimate and Regional Aspects*, edited by A. Berger, F. Mesinger, and D. Šijački, 21–48. Vienna: Springer. https://doi.org/10.1007/978-3-7091-0973-1_2.

Hao, Z., A. AghaKouchak, N. Nakhjiri, and A. Farahmand. 2014. "Global Integrated Drought Monitoring and Prediction System." *Scientific Data* 1: 140001. https://doi.org/10.1038/sdata.2014.1.

Harrabin, R. 2010. "Q&A: Professor Phil Jones." BBC News (website). Accessed February 13, 2018. http://news.bbc.co.uk/2/hi/8511670.stm.

Harrison, D. P. 2017. "Global Negative Emissions Capacity of Ocean Macronutrient Fertilization." *Environmental Research Letters* 12 (3): https://doi.org/10.1088/1748-9326/aa5ef5.

Hartwell, C. A., and D. L. Coursey. 2015. "Revisiting the Environmental Rewards of Economic Freedom." *Economics and Business Letters* 4 (1): 36–50.

Hayek, F. A. 1988. *The Fatal Conceit: The Errors of Socialism.* Chicago: University of Chicago Press.

Hecht, J. 2004. "Corals Adapt to Cope with Global Warming." NewScientist (website). August 11, 2004. https://www.newscientist.com/article/dn6275-corals-adapt-to-cope-with-global-warming/.

Ho, M., and Z. Wang. 2015. "Green Growth for China?" Resources (website). January 14, 2015. https://www.resourcesmag.org/archives/green-growth-for-china/.

Holter, Mikael. 2019. "Norway's Huge New Oil Project Clashes with Growing Focus on Climate." Bloomberg.com (website). October 7, 2019. https://www.bloomberg.com/news/articles/2019-10-07/norway-s-huge-new-oil-project-clashes-with-growing-climate-focus.

Hourdin, F., T. Mauritsen, A. Gettelman, J.-C. Golaz, V. Balaji, Q. Duan, D. Folini, D. Ji, D. Klocke, Y. Qian, F. Rauser, C. Rio, L. Tomassini, M. Watanabe, and D. Williamson 2017. "The Art and Science of Climate Model Tuning." *Bulletin of the American Meteorological Society* 98 (3): 589–602.

Houghton, J. T., G. J. Jenkins, and J. J. Ephraums. 1990. *Climate Change: The IPCC Scientific Assessment.* Cambridge: Cambridge University Press.

Hoyt, D. V., and K. H. Schatten. 1993. "A Discussion of Plausible Solar Irradiance Variations, 1700–1992." *Journal of Geophysical Research* 98 (A11): 18895–906.

Huber, P. 1999. *Hard Green: Saving the Environment from the Environmentalists—A Conservative Manifesto.* New York: Basic Books.

Hughes, G. 2012. "Why Is Wind Power So Expensive? An Economic Analysis." *GWPF Report* 7. London: Global Warming Policy Foundation.

Hurrell, J. W., and H. van Loon. 1994. "A Modulation of the Atmospheric Annual Cycle in the Southern Hemisphere." *Tellus* 46A: 325–38.

Idso, C. D. 2013. *The Positive Externalities of Carbon Dioxide: Estimating the Monetary Benefits of Rising Atmospheric CO2 Concentrations on Global Food Production.* Tempe, AZ: Center for the Study of Carbon Dioxide and Global Change.

Idso, C. D., R. M. Carter, S. F. Singer, and W. Soon. 2013. "Scientific Critique of IPCC's 2013 'Summary for Policymakers.'" *NIPCC Policy Brief.* Nongovernmental International Panel on Climate Change (NIPCC). Arlington Heights, IL: The Heartland Institute.

Idso, S. B. 1989. *Carbon Dioxide and Global Change: Earth in Transition.* Tempe, AZ: IBR Press.

IPPC AR1. 1990. *Climate Change: The IPCC Scientific Assessment.* First Assessment Report of the Intergovernmental Panel on Climate Change. Cambridge: Cambridge University Press.

IPCC AR2. 1996. *Climate Change 1995.* Second Assessment Report of the Intergovernmental Panel on Climate Change. Cambridge: Cambridge University Press.

IPCC AR3. 2001. *Climate Change 2001.* Third Assessment Report of the Intergovernmental Panel on Climate Change. Cambridge: Cambridge University Press.

IPCC AR4. 2007. *Climate Change 2007.* Fourth Assessment Report of the Intergovernmental Panel on Climate Change. Cambridge: Cambridge University Press.

IPCC AR5. 2013, 2014. *Climate Change 2013* (Working Group I contribution) and *Climate Change 2014* (Working Groups II and III contributions). Fifth Assessment Report of the Intergovernmental Panel on Climate Change. Cambridge: Cambridge University Press.

Jacoby, G. C., R. D. D'Arrigo, and T. Davaajamts. 1996. "Mongolian Tree Rings and Twentieth-Century Warming." *Science* 273: 771–3.

Jean, P. 2008. "Al Gore Purchases Carbon Credits From a Company He Himself Owns." Digital Journal (website). March 4, 2008. http://www.digitaljournal.com/article/251232#ixzz6L34shoYX.

JISAO (Joint Institute for the Study of the Atmosphere and Ocean). 2019. "The Pacific Decadal Oscillation." University of Washington (website). Accessed November 4, 2019. http://research.jisao.washington.edu/pdo/.

Kalnay, E., and M. Cai. 2003. "Impact of Urbanization and Land-Use Change on Climate." *Nature* 423: 528–31.

Karl, T. R., A. Arguez, B. Huang, J. H. Lawrimore, J. R. McMahon, M. J. Menne, T. C. Peterson, R. S. Vose, and H.-M. Zhang. 2015. "Possible Artifacts of Data Biases in the Recent Global Surface Warming Hiatus." *Science* 348 (6242): 1469–72.

Karl, T. R., S. J. Hassol, C. D. Miller, and W. L. Murray. 2006. *Temperature Trends in the Lower Atmosphere: Steps for Understanding and Reconciling Differences.* Washington, DC: Climate Change Science Program and the Subcommittee on Global Change Research.

Karl, T. R., and P. D. Jones. 1989. "Urban Bias in Area-Averaged Surface Air Temperature Trends." *Bulletin of the American Meteorological Society* 70: 265–70.

Keigwin, L. D. 1996. "The Little Ice Age and Medieval Warm Period in the Sargasso Sea." *Science* 274: 1504–8.

Kennedy, J., N. A. Rayner, R. O. Smith, D. E. Parker, and M. Saunby 2011. "Reassessing Biases and Other Uncertainties in Sea Surface Temperature Observations Measured in Situ Since 1850: 2. Biases and Homogenization." *Journal of Geophysical Research Atmospheres* 116 (D14): https://doi.org/10.1029/2010JD015220.

Kent, E. C., N. A. Rayner, D. I. Berry, M. Saunby, B. I. Moat, J. J. Kennedy, and D. E. Parker. 2013. "Global Analysis of Night Marine Air Temperature and Its Uncertainty Since 1880: The HadNMAT2 Data Set." *Journal of Geophysical Research Atmospheres* 118 (3): 1281–98.

Kiefer, T. A. 2013. "Energy Insecurity: The False Promise of Liquid Biofuels." *Strategic Studies Quarterly* (Spring 2013): 114–51.

Kleppe, J. A., D. S. Brothers, G. M. Kent, F. Biondi, S. Jensen, and N. W. Driscoll. 2011. "Duration and Severity of Medieval Drought in the Lake Tahoe Basin." *Quaternary Science Reviews* 30: 3269–79.

Klotzbach, P. J., S. G. Bowen, R. Pielke Jr., and M. Bell. 2018. "Continental U.S. Hurricane Landfall Frequency and Associated Damage Observations and Future Risks." *Bulletin of the American Meteorological Society* 99 (7): 1359–77.

Klotzbach, P. J., J. C. L. Chan, P. J. Fitzpatrick, W. M. Frank, C. W. Landsea, and J. L. McBride. 2017. "The Science of William M. Gray: His Contributions to the Knowledge of Tropical Meteorology and Tropical Cyclones." *Bulletin of the American Meteorological Society* 98: 2311–36.

Kollmuss, A., L. Schneider, and V. Zhezherin. 2015. "Has Joint Implementation Reduced GHG Emissions? Lessons Learned for the Design of Carbon Market Mechanisms." *SEI Working Paper* No. 2015-07. Stockholm: Stockholm Environment Institute.

Kummer, L. 2015. "The 97% Consensus of Climate Scientists Is Only 47%." Fabius Maximus (website). https://fabiusmaximus.com/2015/07/29/new-study-undercuts -ipcc-keynote-finding-87796/.

Lamb, H. H. 1985. *Climatic History and the Future*. Princeton, NJ: Princeton University Press.

Landsea, C. 2005. "Resignation Letter of Chris Landsea from IPCC." Climatechange-facts.info (website). Accessed November 20, 2018. http://www.climatechangefacts. info/ClimateChangeDocuments/LandseaResignationLetterFromIPCC.htm.

Laster, H., A. M. Lenchek, and S. F. Singer. 1962. "Forbush Decreases Produced by Diffusive Deceleration Mechanism in Interplanetary Space." *Journal of Geophysical Research* 67 (7): 2639–43.

Lewis, N. 2013. "An Objective Bayesian, Improved Approach for Applying Optimal Fingerprint Techniques to Estimate Climate Sensitivity." *Journal of Climate.* https:// doi.org/10.1175/JCLI-D-12-00473.1.

Ligtenberg, S. R. M., W. J. Berg, M. R. Broeke, J. G. L. Rae, and E. Meijgaard. 2013. "Future Surface Mass Balance of the Antarctic Ice Sheet and Its Influence on Sea Level Change, Simulated by a Regional Atmospheric Climate Model." *Climate Dynamics* 41: 867–84.

Lindzen, R. S. 2012. "Climate Science: Is It Currently Designed to Answer Questions?" *Euresis Journal* 2 (Winter): 161–92.

——. 2015. "Global Warming, Models and Language." In *Climate Change: The Facts*, edited by A. Moran, 38–56. Melbourne, Australia and Woodville, NH: Institute of Public Affairs and Stockade Books.

Lindzen, R. S., and Y-S. Choi. 2011. "On the Observational Determination of Climate Sensitivity and Its Implications." *Asia-Pacific Journal of Atmospheric Science* 47: 377–90.

Loaiciga, H. A. 2006. "Modern-Age Buildup of CO_2 and Its Effects on Seawater Acidity and Salinity." *Geophysical Research Letters* 33: L10605, https://doi.org/10.1029/2006GL026305.

Liu, Y., W. Liu, Z. Peng, Y. Xiao, G. Wei, W. Sun, J. He, G. Liu, and C.-L. Chou. 2009. "Instability of Seawater pH in the South China Sea during the Mid-late Holocene: Evidence from Boron Isotopic Composition of Corals." *Geochimica et Cosmochimica Acta* 73: 1264–72.

Maibach, E., D. Perkins, K. Timm, T. Myers, B. Woods Placky, S. Sublette, A. Engblom, and K. Seitter. 2017. *A 2017 National Survey of Broadcast Meteorologists: Initial Findings*. Fairfax, VA: George Mason University, Center for Climate Change Communication.

Malone, T. F., ed. 1951. *Compendium of Meteorology*. Boston: The American Meteorological Society.

Mann, M. E., R. S. Bradley, and M. K. Hughes. 1998. "Global-Scale Temperature Climate Forcing over the Past Six Centuries." *Nature* 392: 779–87.

——. 1999. "Northern Hemisphere Temperatures During the Past Millennium: Inferences, Uncertainties, and Limitations." *Geophysical Research Letters* 26: 759–62.

——. 2004. "Correction: Corrigendum: Global-Scale Temperature Patterns and Climate Forcing over the Past Six Centuries." *Nature* 430: 105.

Masood, E. 1996. "Sparks Fly over Climate Report." *Nature* 381: 639.

Maue, R. 2020. Global Tropical Cyclone Activity (website). Accessed August 2, 2020. http://climatlas.com/tropical/.

Mayo, J. W., and J. E. Mathis. 1988. "The Effectiveness of Mandatory Fuel Efficiency Standards in Reducing the Demand for Gasoline." *Applied Economics* 20: 211–9.

McCulloch, M. T., J. P. D'Olivo, J. Falter, M. Holcomb, and J. A. Trotter. 2017. "Coral Calcification in a Changing World and the Interactive Dynamics of pH and DIC Upregulation." *Nature Communications* 8: 15686.

McIntyre, S., and R. McKitrick. 2003. "Corrections to the Mann et al. (1998) Proxy Data Base and Northern Hemispheric Average Temperature Series." *Energy & Environment* 14: 751–77.

——. 2005. "Hockey Sticks, Principal Components and Spurious Significance." *Geophysical Research Letters* 32: 3.

McKitrick, R. 2001. "Mitigation Versus Compensation in Global Warming Policy." *Economics Bulletin* 17 (2): 1–6.

McKitrick, R., and J. Christy. 2018. "A Test of the Tropical 200- to 300-hPa Warming Rate in Climate Models." *Earth and Space Science* 5: 529–36.

McKitrick, R., and P. J. Michaels. 2007. "Quantifying the Influence of Anthropogenic Surface Processes and Inhomogeneities on Gridded Global Climate Data." *Journal of Geophysical Research* 112 (24). https://doi:10.1029/2007JD008465.

McLean, J. 2018. *An Audit of the Creation and Content of the HadCRUT4 Temperature Dataset*. Beaverton, OR: Robert Boyle Publishing.

Meadows, D. H., D. L. Meadows, J. Randers, and W. W. Behrens III. 1972. *Limits to Growth*. New York: Universe Books.

Michaels, P. 2009. "The Dog Ate Global Warming." National Review Online (website). September 23, 2009. https://www.nationalreview.com/2009/09/dog-ate-global-warming-patrick-j-michaels/.

Miller, A. J., D. R. Cayan, T. P. Barnett, N. E. Graham, and J. M. Oberhuber. 1994. "The 1976–77 Climate Shift of the Pacific Ocean." *Oceanography* 7 (1): 21–6.

Minnis, P., J. K. Ayers, R. Palikonda, and D. Phan. 2004. "Contrails, Cirrus Trends, and Climate." *Journal of Climate* 17: 1671–85.

Minx, J. C., W. F. Lamb, M. W. Callaghan, L. Bornmann, and S. Fuss. 2017. "Fast Growing Research on Negative Emissions." *Environmental Research Letters* 12 (3). https://doi.org/10.1088/1748-9326/aa5ee5.

Mitchell Jr., J. M. 1970. "A Preliminary Evaluation of Atmospheric Pollution as a Cause of the Global Temperature Fluctuation of the Past Century." In *Global Effects of Environmental Pollution: A Symposium Organized by the American Association for the Advancement of Science* held in Dallas, Texas, December 1968, edited by S. F. Singer, 139–55. Berlin: Springer.

Monckton, C. 2016. "Is CO_2 Mitigation Cost Effective?" In *Evidence-Based Climate Science*, 2nd ed., edited by D. Easterbrook, 175–87. Amsterdam: Elsevier.

Monckton, C., W. W.-H. Soon, D. R. Legates, and W. M. Briggs. 2015. "Why Models Run Hot: Results from an Irreducibly Simple Climate Model." *Science Bulletin* 60 (1): 122–35.

Montford, A. W. 2010. *The Hockey Stick Illusion: Climategate and the Corruption of Science*. London: Stacey International.

Mooney, K. 2016. "House Probe Reveals Audit Detailing Climate Change Researcher's 'Double Dipping.'" Daily Signal (website). March 2, 2016. https://kevinmooney.info/2016/03/house-probe-reveals-audit-detailing-climate-change-researchers-double-dipping/.

Moore, S., and J. Simon. 2000. *It's Getting Better All the Time: 100 Greatest Trends of the Last 100 Years*. Washington, DC: Cato Institute.

Moore, S., and K. White. 2016. *Fueling Freedom: Exposing the Mad War on Energy*. Washington, DC: Regnery Publishing.

Moore, T. G. 1995. *Global Warming: A Boon to Humans and Other Animals*. Stanford, CA: Hoover Institution Press.

Murdock, C. C., E. D. Sternberg, and M. B. Thomas. 2016. "Malaria Transmission Potential Could Be Reduced with Current and Future Climate Change." *Scientific Reports* 6: https://doi.org/10.1038/srep27771.

Nafstad, P., A. Skrondal, and E. Bjertness. 2001. "Mortality and Temperature in Oslo, Norway, 1990–1995." *European Journal of Epidemiology* 17: 621–7.

NAS (National Academy of Sciences). 2006. *Surface Temperature Reconstructions for the Last 2,000 Years*. Washington, DC: The National Academies Press.

NASA/GISS (National Aeronautics and Space Administration, Goddard Institute for Space Studies). 2019. "Global Mean Estimates of Land and Ocean Data." National Aeronautics and Space Administration, Goddard Institute for Space Studies (website). Accessed March 17, 2019. https://data.giss.nasa.gov/gistemp/graphs/.

National Assessment Synthesis Team. 2001. *Climate Change Impacts on the United States: The Potential Consequences of Climate Variability and Change*. Report for the US Global Change Research Program. Cambridge: Cambridge University Press.

National Centers for Environmental Information. "Global Historical Climatology Network (GHCN)." National Climatic Data Center (website). Accessed March 17, 2019. https://www.ncdc.noaa.gov/data-access/land-based-station-data/land-based-datasets/global-historical-climatology-network-ghcn.

Nature. 1996. "Climate Debate Must Not Overheat." Editorial. *Nature* 381: 539.

Nichols, N. 2013. "The Tax-Exemption Rip-Off." Townhall (website). October 1, 2013. https://townhall.com/columnists/-nicknichols/2013/10/01/the-taxexemption-ripoff-n1713419.

NIPCC (Nongovernmental International Panel on Climate Change). 2009. *Climate Change Reconsidered: The 2009 Report of the Nongovernmental International Panel on Climate Change*, edited by C. D. Idso and S. F. Singer. Chicago: The Heartland Institute.

———. 2011. *Climate Change Reconsidered: 2011 Interim Report*, edited by C. D. Idso, R. M. Carter, and S. F. Singer. Chicago: The Heartland Institute.

———. 2013. *Climate Change Reconsidered: Physical Science*, edited by C. D. Idso, R. M. Carter, and S. F. Singer. Chicago: The Heartland Institute.

———. 2014. *Climate Change Reconsidered II: Biological Impacts*, edited by C. D. Idso, S. B. Idso, R. M. Carter, and S. F. Singer. Chicago: The Heartland Institute.

———. 2019. *Climate Change Reconsidered II: Fossil Fuels*, edited by R. Bezdek, C. D. Idso, D. Legates, and S. F. Singer. Arlington Heights, IL: The Heartland Institute.

Nordhaus, T., and M. Shellenberger. 2007. *Break Through: From the Death of Environmentalism to the Politics of Possibility.* Boston: Houghton Mifflin Company.

Nordhaus, W. D. 1991. "The Cost of Slowing Down Climate Change: A Survey." *Energy Journal* 12 (1): 37–65.

NRC (National Research Council). 2000. *Reconciling Observations of Global Temperature Change.* Washington, DC: The National Academies Press.

Oerlemans, J. 1982. "Response of the Antarctic Ice Sheet to Climate Warming." *Journal of Climate* 2: 1–12.

OMB (US Office of Management and Budget). 2013. *Draft Report to Congress on the Benefits and Costs of Federal Regulations and Agency Compliance with the Unfunded Mandates Reform Act.* Washington, DC: Office of Management and Budget.

Ostrom, E. 2010. "Polycentric Systems for Coping with Collective Action and Global Environmental Change." *Global Environmental Change* 20: 550–57.

Panayotou, T. 1993. "Empirical Tests and Policy Analysis of Environmental Degradation at Different Stages of Economic Development." *Working Paper* WP238. Geneva: International Labour Office, Technology and Employment Programme.

Parker, A., and C. D. Ollier. 2016. "Coastal Planning Should Be Based on Proven Sea Level Data." *Ocean & Coastal Management* 124: 1–9.

———. 2017. "Short-Term Tide Gauge Records from One Location Are Inadequate to Infer Global Sea Level Acceleration." *Earth Systems and Environment* 1 (17). https://doi.org/10.1007/s41748-017-0019-5.

Parry, M., N. Arnell, M. Hulme, R. Nicholls, and M. Livermore. 1998. "Adapting to the Inevitable." *Nature* 395: 741.

Peterson, T. C., and R. S. Vose. 1997. "An Overview of the Global Historical Climatology Network Temperature Database." *Bulletin of the American Meteorological Society* 78 (12): 2837–49.

Pielke Jr., R. A., C. Landsea, M. Mayfield, J. Laver, and R. Pasch. 2005. "Hurricanes and Global Warming." *Bulletin of the American Meteorological Society* 86: 1571–5.

Pielke Sr., R. A., C. A. Davey, D. Niyogi, S. Fall, J. Steinweg-Woods, K. Hubbard, X. Lin, M. Cai, Y.-K. Lim, H. Li, J. Nielsen-Gammon, K. Gallo, R. Hale, R. Mahmood, S. Foster, R. T. McNider, and P. Blanken. 2007a. "Unresolved Issues with the Assessment of Multidecadal Global Land Surface Temperature Trends." *Journal of Geophysical Research* 112 (D24). https://doi.org/10.1029/2006JD008229.

Pielke Sr., R. A., J. Nielsen-Gammon, C. Davey, J. Angel, O. Bliss, N. Doesken, M. Cai, S. Fall, D. Niyogi, K. Gallo, R. Hale, K. G. Hubbard, X. Lin, J. Li, and S. Raman. 2007b. "Documentation of Uncertainties and Biases Associated with Surface Temperature Measurement Sites for Climate Change Assessment." *Bulletin of the American Meteorological Society* 88: 913–28.

Pindyck, R. S. 2013. "Climate Change Policy: What Do the Models Tell Us?" *Journal of Economic Literature* 51: 860–72.

Plass, G. N. 1956. "The Carbon Dioxide Theory of Climatic Change." *Tellus* 8 (2): 140–54.

Rahmstorf, S. 2007. "A Semi-empirical Approach to Projecting Future Sea Level Rise." *Science* 315 (5810): 368–70.

Revelle, R. 1977. "Let the Waters Bring Forth Abundantly." In *Arid Zone Development, Potentialities and Problems*, edited by Y. Mundlak and S. F. Singer, 191–200. Cambridge, MA: Ballinger Publishing Company.

Revelle, R., and H. E. Suess. 1957. "Carbon Dioxide Exchange Between Atmosphere and Ocean and the Question of an Increase of Atmospheric CO_2 During the Past Decades." *Tellus* 9: 18–27.

Ring, M. J., D. Lindner, E. F. Cross, and M. E. Schlesinger. 2012. "Causes of the Global Warming Observed Since the 19th Century." *Atmospheric and Climate Sciences* 2: 401–15.

Roberts, D. 2006. "An Interview with Accidental Movie Star Al Gore." Grist Magazine (website). May 9, 2006. https://grist.org/article/roberts2/.

Rusticucci, M. 2012. "Observed and Simulated Variability of Extreme Temperature Events over South America." *Atmospheric Research* 106: 1–17.

Santer, B. D., P. W. Thorne, L. Haimberger, K. E. Taylor, T. M. L. Wigley, J. R. Lanzante, S. Solomon, M. Free, P. J. Gleckler, P. D. Jones, T. R. Karl, S. A. Klein, C. Mears, D. Nychka, G. A. Schmidt, S. C. Sherwood, and F. J. Wentz. 2008. "Consistency of Modelled and Observed Temperature Trends in the Tropical Troposphere." *International Journal of Climatology* 28 (13): 1703–22.

Scafetta, N., and B. J. West. 2006. "Phenomenological Solar Contribution to the 1900–2000 Global Surface Warming." *Geophysical Research Letters* 33: L05708, https://doi.org/10.1029/2005GL025539.

Scafetta, N., and R. C. Willson. 2014. "ACRIM Total Solar Irradiance Satellite Composite Validation versus TSI Proxy Models." *Astrophysical Space Science* 350: 421–2.

Schell, J. 1989. "Our Fragile Earth." *Discover* (October): 45–8.

Schmithüsen, H., J. Notholt, G. König-Langlo, P. Lemke, and T. Jung. 2015. "How Increasing CO_2 Leads to an Increased Negative Greenhouse Effect in Antarctica." *Geophysical Research Letters* 42: 23.

Schneider, S. H. 1976. *The Genesis Strategy: Climate and Global Survival.* New York: Plenum Press.

———. 1989. *Global Warming: Are We Entering the Greenhouse Century?* San Francisco: Sierra Club Books.

Schreier, M., A. Kokhanovsky, V. Eyring, L. Bugliaro, H. Mannstein, B. Mayer, H. Bovensmann, and J. P. Burrows. 2006. "Impact of Ship Emissions on the Microphysical, Optical and Radiative Properties of Marine Stratus: A Case Study." *Atmospheric Chemistry and Physics* 6: 4925–42.

Seager, R., N. Graham, C. Herweijer, A. L. Gordon, Y. Kushnir, and E. Cook 2007. "Blueprints for Medieval Hydroclimate." *Quaternary Science Reviews* 26: 2322–36.

Seitz, F. 1996. "A Major Deception on Global Warming." *Wall Street Journal.* June 12, 1996.

Shepherd, J. G. 2009. *Geoengineering the Climate: Science, Governance and Uncertainty.* Working Group on Geoengineering the Climate. RS Policy Document 10/29. London: Royal Society.

Simon, J. L., ed. 1995. *The State of Humanity.* Oxford: Blackwell Publishers.

Singer, S. F. 1979. "Cost Benefit Analysis as an Aid to Environmental Decision Making." McLean, VA: The MITRE Corporation.

———. 1988. "Re-analysis of 'Nuclear Winter.'" *Meteorology & Atmospheric Physics* 38: 228–39.

———. 1989. *Global Climate Change: Human and Natural Influences.* New York: Paragon House Publishers.

———. 1997a. "Climate Warming from Increasing Air Traffic?" Paper presented at NASA Conference on the Atmospheric Effects of Aviation, Virginia Beach, VA, March 10–14, 1997.

———. 1997b. "Global Warming Will Not Raise Sea-Levels." Abstract. 1997 Fall Meeting of the American Geophysical Union, San Francisco, CA, December 8–12, 1997.

———. 1998. "Ironies Grown in Kyoto." *Washington Times.* January 23, 1998, A19.

———. 1999. "Human Contribution to Climate Change Remains Questionable." *Eos* 80: 16.

———. 2000. "Climate Policy—from Rio to Kyoto: A Political Issue for 2000—and Beyond." *Essays in Public Policy* 102. Stanford, CA: Hoover Institution Press.

———. 2003. "The Revelle-Gore Story: Attempted Political Suppression of Science." In Gough, M., ed. *Politicizing Science: The Alchemy of Policymaking.* Stanford, CA: Hoover Institution Press.

———. 2009. "Climategate: Some Comments by Prof. S. Fred Singer." Institute of Economic Affairs (website). November 25, 2009. https://iea.org.uk/blog/climategate -some-comments-by-prof-s-fred-singer.

———. 2011. "Lack of Consistency Between Modeled and Observed Temperature Trends." *Energy & Environment* 22 (4): 375–406.

———. 2013. "Inconsistency of Modeled and Observed Tropical Temperature Trends." *Energy & Environment* 24 (3–4): 405–13.

———. 2015. "Paris Climate Conference Is Likely to Fail." *American Thinker* (website). October 22, 2015. https://www.americanthinker.com/articles/2015/10/paris_climate_conference_is_likely_to_fail.html.

———. 2018. "Making Climate Predictions." *Washington Times.* November 28, 2018.

Singer, F. S., and D. T. Avery. 2008. *Unstoppable Global Warming: Every 1,500 Years.* Updated and expanded ed. Lanham, MD: Rowman and Littlefield Publishers, Inc.

Singer, F. S., B. A. Boe, F. W. Decker, N. Frank, T. Gold, W. Gray, H. Linden, R. Lindzen, P. J. Michaels, W. A. Nierenberg, and R. Stevenson. 1997. "Comments on 'Open Letter to Ben Santer.'" *Bulletin of the American Meteorological Society* 78 (1): 81–2.

Singer, F. S., R. Revelle, and C. Starr. 1991. "What to Do About Global Warming: Look Before You Leap." *Cosmos* 1 (1): 28–33.

Slade, R., R. Saunders, R. Gross, and A. Bauen. 2011. *Energy from Biomass: The Size of the Global Resource.* London: Imperial College Centre for Energy Policy and Technology and UK Energy Research Centre.

SMIC (Study of Man's Impact on Climate). 1971. *Inadvertent Climate Modification: Report of the Study of Man's Impact on Climate,* edited by C. L. Wilson and W. H. Matthews. Cambridge, MA: MIT Press.

Song, X., S. Wang, T. Li, J. Tian, G. Ding, J. Wang, and K. Shang. 2018. "The Impact of Heat Waves and Cold Spells on Respiratory Emergency Department Visits in Beijing, China." *Science of the Total Environment* 615: 1499–1505.

Soon, W. W.-H. 2005. "Variable Solar Irradiance as a Plausible Agent for Multidecadal Variations in the Arctic-Wide Surface Air Temperature Record of the Past 130 Years." *Geophysical Research Letters* 32. https://doi.org/10.1029/2005GL023429.

———. 2009. "Solar Arctic-Mediated Climate Variation on Multidecadal to Centennial Timescales: Empirical Evidence, Mechanistic Explanation and Testable Consequences." *Physical Geography* 30: 144–84.

Soon, W., S. L. Baliunas, C. D. Idso, S. Idso, and D. R. Legates. 2003. "Reconstructing Climatic and Environmental Changes of the Past 1000 Years: A Reappraisal." *Energy & Environment* 14 (2/3): 233–96.

Soon, W., R. Connolly, and M. Connolly. 2015. "Re-evaluating the Role of Solar Variability on Northern Hemisphere Temperature Trends Since the 19th Century." *Earth-Science Reviews* 150: 409–52.

Soon, W., D. R. Legates, and S. L. Baliunas. 2004. "Estimation and Representation of Long-Term (>40 year) Trends of Northern-Hemisphere-gridded Surface Temperature: A Note of Caution." *Geophysical Research Letters* 31. https://doi:10.1029/2003GL019141.

Soon, W. W.-H., E. S. Posmentier, and S. L. Baliunas. 1996. "Inference of Solar Irradiance Variability from Terrestrial Temperature Changes, 1880–1993." *Astrophysical Journal* 472: 891–902.

Spencer, R. 2018. "UAH Global Temperature Update for October." DrRoySpencer. com (blog). November 2, 2018. https://www.drroyspencer.com/2018/11/uah-global -temperature-update-for-october-2018-0-22-deg-c/.

———. 2019. "UAH Global Temperature Update for October." Dr.RoySpencer.com (blog). November 1, 2019. https://www.drroyspencer.com/2019/11/uah-global -temperature-update-for-october-2019-0-46-deg-c/.

Stacy, T. F., and G. S. Taylor. 2016. *The Levelized Cost of Electricity from Existing Generation Resources.* Washington, DC: Institute for Energy Research.

Stager, J. C., and P. A. Mayewski. 1997. "Abrupt Early to Mid-Holocene Climatic Transition Registered at the Equator and the Poles." *Science* 276: 1834–6.

Stavins, R. 2014. "Is the IPCC Government Approval Process Broken?" An Economic View of the Environment (blog). April 25, 2014. http://www.robertstavinsblog. org/2014/04/25/is-the-ipcc-government-approval-process-broken-2/.

Stott, R. 2012. "Contraction and Convergence: The Best Possible Solution to the Twin Problems of Climate Change and Inequity." *The BMJ* 344. https://doi.org/10.1136/ bmj.e1765.

Svensmark, H., M. B. Enghoff, N. J. Shaviv, and J. Svensmark. 2017. "Increased Ionization Supports Growth of Aerosols into Cloud Condensation Nuclei." *Nature Communications* 8: 2199.

Tans, P. 2009. "An Accounting of the Observed Increase in Oceanic and Atmospheric CO_2 and an Outlook for the Future." *Oceanography* 22: 26–35.

Tol, R. 2014. "IPCC Again." Richard Tol Occasional Thoughts on All Sorts (blog). April 25, 2014. http://richardtol.blogspot.com/2014/04/ipcc-again.html.

Tol, R., and S. Wagner. 2010. "Climate Change and Violent Conflict in Europe over the Last Millennium." *Climatic Change* 99: 65–79.

Toscano, M. A., and I. G. Macintyre. 2003. "Corrected Western Atlantic Sea-Level Curve for the Last 11,000 Years Based on Calibrated 14 C Dates from Acropora." *Coral Reefs* 22: 257–70.

Travis, D. J., A. M. Carleton, and R. G. Lauritsen. 2012. "Contrails Reduce Daily Temperature Range." *Nature* 418 (8): 601.

Trenberth, K. E., and J. W. Hurrell. 1994. "Decadal Atmosphere-Ocean Variations in the Pacific." *Climate Dynamics* 9: 303–19.

Trenberth, K., and J. T. Fasullo. 2013. "An Apparent Hiatus in Global Warming?" *Earth's Future* 1: 19–32.

US Census Bureau. 2016. *An Aging World: 2015.* International Population Reports P95/16-1. Washington, DC: US Department of Commerce.

US Global Change Research Program. 2019. "Sea Level Rise." GlobalChange.gov (website). Accessed June 29. 2019. https://www.globalchange.gov/browse/indicators/global-sea-level-rise.

van Wijngaarden, W. A., and W. Happer. 2018. "Effect of Linewidth Narrowing on Radiative Forcing." Paper presented at American Meteorological Society 15th Conference on Atmospheric Radiation, Vancouver, British Columbia, July 9–13, 2018.

Voosen, P. 2016. "Climate Scientists Open Up Their Black Boxes to Scrutiny." *Science* 354: 401–2.

Wang, H., J. Chen, S. Zhang, D. D. Zhang, Z. Wang, Q. Xu, S. Chen, S. Wang, S. Kang, and F. Chen. 2018. "A Chironomid-Based Record of Temperature Variability During the Past 4000 years in Northern China and Its Possible Societal Implications." *Climate of the Past* 14: 383–96.

Wanning, E. 1984. "Interview: Roger Revelle." *Omni* 6 (March): 77–83, 112.

Weart, S. R. 1997. "The Discovery of the Risk of Global Warming." *Physics Today* 50: 34–40.

Wegman, E., D. W. Scott, and Y. Said. 2006. *Ad Hoc Committee Report to Chairman of the House Committee on Energy & Commerce and to the Chairman of the House Subcommittee on Oversight & Investigations on the Hockey-stick Global Climate Reconstructions.* Washington, DC: US House of Representatives.

Wei, Z., X. Fang, and Y. Su. 2015. "A Preliminary Analysis of Economic Fluctuations and Climate Changes in China from BC 220 to AD 1910." *Regional Environmental Change* 15: 1773–85.

Weitzman, M. L. 2015. "Review of *The Climate Casino: Risk, Uncertainty, and Economics for a Warming World*, by William Nordhaus." *Review of Environmental Economics and Policy* 9: 145–56.

White, C. 2017. "The Dynamic Relationship Between Temperature and Morbidity." *Journal of the Association of Environmental and Resource Economists* 4: 1155–98.

Wigley, T. M. L., and S. C. B. Raper. 1992. "Implications on Climate and Sea Level of Revised IPCC Emissions Scenarios." *Nature* 357: 293–300.

Wood, Peter. 2011. "Scholars Critique Campus 'Sustainability' Movement, Propose Alternatives." News release. National Association of Scholars, April 21, 2011.

Wylie, R. 2013. "Long Invisible, Research Shows Volcanic CO2 Levels Are Staggering." Live Science (website). October 15, 2013. https://www.livescience.com/40451-volcanic-co2-levels-are-staggering.html.

Zhang, X., F. W. Zwiers, G. Li, H. Wan, and A. J. Cannon. 2017. "Complexity in Estimating Past and Future Extreme Short-Duration Rainfall." *Nature Geoscience* 10: 255–59.

Zhu, Z., S. Piao, Z. Zhu, S. Piao, R. B. Myneni, M. Huang, Z. Zeng, J. G. Canadell, P. Ciais, S. Sitch, P. Friedlingstein, A. Arneth, C. Cao, L. Cheng, E. Kato, C. Koven, Y. Li, X. Lian, Y. Liu, R. Liu, J. Mao, Y. Pan, S. Peng, J. Peñuelas, B. Poulter, T. A. M. Pugh, B. D. Stocker, N. Viovy, X. Wang, Y. Wang, Z. Xiao, H. Yang, S. Zaehle, and N. Zeng. 2016. "Greening of the Earth and Its Drivers." *Nature Climate Change* 6: 791–95.

Ziskin, S., and N. J. Shaviv. 2012. "Quantifying the Role of Solar Radiative Forcing over the Twentieth Century." *Advances in Space Research* 50 (6): 762–76.

Acronyms

ACE	accumulated cyclone energy
ACRIM	Active Cavity Radiometer Irradiance Monitor satellite
AGW	anthropogenic global warming
AIRS	atmospheric infrared sounder
ALR	atmospheric lapse rate
AMOC	Atlantic Meridional Overturning Circulation
AMS	American Meteorological Society
AR1	First Assessment Report of the IPCC (1990)
AR2	Second Assessment Report of the IPCC (1996)
AR3	Third Assessment Report of the IPCC (2001)
AR4	Fourth Assessment Report of the IPCC (2007)
AR5	Fifth Assessment Report of the IPCC (2013, 2014)
BASIC	Brazil, South Africa, India, and China
BEST	Berkeley (CA) Earth System Temperature
C	Celsius
C&C	contraction and convergence
CBA	cost-benefit analysis

CDM	clean development mechanism
CEL	characteristic emission layer
CERN	European Institute for Nuclear Research
CFC	chlorofluorocarbon
CH4	methane
cm	centimeter
CMIP	Coupled Model Intercomparison Project
CO2	carbon dioxide
COP	Conference of the Parties
CPP	Clean Power Plan
CRU	Climatic Research Unit at the University of East Anglia
DOB	Dansgaard-Oeschger Bond
ECS	equilibrium climate sensitivity
EIA	Energy Information Administration (USA)
EKC	environmental Kuznets curve
ENSO	El Niño/Southern Oscillation
EPA	Environmental Protection Agency (USA)
EPRF	Energy Policy Research Foundation
EPRI	Electric Power Research Institute
ERI	engine room inlet
F	Fahrenheit
FAO	Food and Agriculture Organization of the United Nations
FCCC	United Nations Framework Convention on Climate Change

FERC	Federal Energy Regulatory Commission (USA)
GCM	general circulation model or global climate model
GDP	gross domestic product
GHCN	Global Historical Climatology Network
GHG	greenhouse gas
GHW	greenhouse warming
GIMMS	global inventory modeling and mapping studies
GISS	Goddard Institute for Space Studies
GRIP	Greenland Ice Core Project
H2O	hydrogen dioxide (water)
HadAT	Hadley Centre dataset of global radiosonde gridded temperature anomalies
HadCRUT	Hadley Centre/Climatic Research Unit dataset of monthly instrumental temperature records
hPa	hectopascal pressure unit
IPCC	United Nations' Intergovernmental Panel on Climate Change
IR	infrared
LAI	leaf area index
LIA	Little Ice Age
LRSL	local relative sea level
MIT	Massachusetts Institute of Technology
MSU	microwave sounding unit
MWP	Medieval Warm Period
N2O	nitrous oxide

NAO	North Atlantic Oscillation
NAS	National Academy of Sciences (USA)
NASA	National Aeronautics and Space Administration (USA)
NCAR	National Center for Atmospheric Research (USA)
NDC	nationally determined contribution
NET	negative emission technology
NEXTRAD	next-generation radar
NGO	nongovernmental organization
NH	Northern Hemisphere
NIPCC	Nongovernmental International Panel on Climate Change
NMAT	nighttime marine air temperature
NOAA	National Oceanic and Atmospheric Administration (USA)
NPP	net primary productivity (production)
O_3	ozone
OLR	outgoing long-wave radiation
PDO	Pacific Decadal Oscillation
Pg	petagrams
ppm	parts per million
RATPAC	Radiosonde Atmospheric Temperature Products for Assessing Climate
RCP	representative concentration pathway
RF	radiative forcing
SD	sustainable development
SEPP	Science and Environmental Policy Project

SL	sea level
SMIC	Study of Man's Impact on Climate
SPM	Summary for Policymakers
SST	sea surface temperature
TCR	transient climate response
TMT	tropical midtropospheric temperature
TOA	top of atmosphere
TSI	total solar irradiance
UAH	University of Alabama–Huntsville
UCAR	University Corporation for Atmospheric Research
UN	United Nations
UNEP	United Nations Environment Programme
UN FCCC	United Nations Framework Convention on Climate Change
UT	upper troposphere
WGI	Working Group I of IPCC
WGII	Working Group II of IPCC
WGIII	Working Group III of IPCC
W/m²	watts per square meter
WMO	World Meteorological Organization
WV	water vapor

Index

About the Authors

S. FRED SINGER (1924-2020) was one of the world's preeminent authorities on energy and environmental issues. A pioneer in the development of rocket and satellite technology, Dr. Singer designed the first satellite instrument for measuring atmospheric ozone and was a principal developer of scientific and weather satellites. He was a research fellow at the Independent Institute, founder and chairman emeritus of the Science and Environmental Policy Project (SEPP), founder of the Nongovernmental International Panel on Climate Change (NIPCC), and professor emeritus at the University of Virginia, where he taught from 1971 to 1994. Dr. Singer received his PhD in physics from Princeton University.

Dr. Singer was the recipient of the White House Special Commendation, Gold Medal Award from the US Department of Commerce, and first Science Award from the British Interplanetary Society. A fellow of the American Association for the Advancement of Science, he received an honorary doctorate from Ohio State University and was elected to the International Academy of Astronautics.

Dr. Singer served as vice chairman of the National Advisory Committee on Oceans and Atmospheres; chief scientist for the US Department of Transportation; deputy assistant administrator at the US Environmental Protection Agency; deputy assistant secretary at the US Department of the Interior; first dean of the School of Environmental and Planetary Sciences, University of Miami; first director of the US Weather Satellite Center (Department of Commerce); director of the Center for Atmospheric and Space Physics, University of Maryland; and research physicist, Upper Atmospheric Rocket Program, Johns Hopkins University.

Dr. Singer also was a visiting scholar at the Woodrow Wilson International Center for Scholars; Jet Propulsion Laboratory, California Institute of Technology; National Air and Space Museum; Lyndon Baines Johnson School for Public Affairs, University of Texas; George Mason University; and the Soviet Academy of Sciences Institute for Physics of the Earth.

Dr. Singer was the author, coauthor, or editor of many books, including *The Changing Global Environment, Free Market Energy, Global Climate Change, Is There an Optimum Level of Population?, Unstoppable Global Warming: Every 1,500 Years*, and five volumes in the *Climate Change Reconsidered* series. He was also the author of more than 400 technical articles in scientific, economics, and public policy journals plus more than 400 articles in popular publications including the *Wall Street Journal, New York Times, Washington Post*, and online at *American Thinker*.

DAVID R. LEGATES is a research fellow at the Independent Institute, Deputy Assistant Secretary of Commerce for Observation and Prediction, Executive Director of the United States Global Change Research Program, and Professor of Climatology in the Department of Geography and an Adjunct Professor in the Department of Applied Economics and Statistics at the University of Delaware. He received his PhD in climatology from the University of Delaware and has taught at Louisiana State University, University of Oklahoma, and University of Virginia. He has been research scientist at the Southern Regional Climate Center, chief research scientist at the Center for Computational Geosciences, and visiting research scientist at the National Climate Data Center. Presently, he serves as the Deputy Assistant Secretary of Commerce for Environmental Observation and Prediction.

Dr. Legates has been published more than 125 times in refereed journals, conference proceedings, and monograph series and has made more than 250 professional presentations. His research has appeared in such journals as the *International Journal of Climatology, The Professional Geographer, Journal of Geophysical Research, Journal of Environmental Hydrology, Journal of Climate, Climatic Change, Bulletin of the American Meteorological Society, Water Resources Bulletin, Geographical Review, Global and Planetary Change, Journal of Hydrology, Theoretical and Applied Climatology, The American Cartographer,* and *Journal of Geophysical Research.* He is a contributor to the *Encyclopedia of Climate and Weather* and *Yearbook of Science and Technology.*

Dr. Legates has argued for the necessity of technological progress in precipitation measurement used for validating climate change scenarios and for validation of existing data used for that purpose. He codeveloped methods to correct bias in gauge-measured precipitation data for wind and temperature effects, with direct applicability in climate change, hydrology, and environmental impact studies. Dr. Legates also developed a calibration method that validates NEXRAD radar precipitation data with gauge measurements to improve the accuracy of precipitation estimates.

Dr. Legates has earned certified consulting meteorologist status from the American Meteorological Society and in 1999 was awarded the Boeing Autometric Award for submitting the best paper in image analysis and interpretation. At the 10th International Conference on Climate Change in 2015, he was presented with the Courage in Defense of Science Award.

ANTHONY R. LUPO is a research fellow at the Independent Institute and professor of Atmospheric Science, and principal investigator of the Global Climate Change Group in the School of Natural Resources at the University of Missouri. He received his PhD in atmosphere science from Purdue University, and he has been a member of the Working Groups I and III for the Intergovernmental Panel on Climate Change (IPCC), research scholar at the Belgorod University in Russia, Fulbright Research Scholar at the Russian Academy of Sciences, associate editor of the *Monthly Weather Review*, and postdoctoral research associate in the Department of Earth and Atmospheric Sciences at the University at Albany.

Dr. Lupo is the author of 135 peer-reviewed publications including in such scientific journals as *Advances in Environmental Biology; Advances in Meteorology; Atmofera; Atmosphere; Atmospheric and Climate Science; Atmospheric Environment; Atmospheric Research; Atmospheric Science Letters; Bulletin of the American Meteorological Society; Climate Dynamics; Current Climate Change Reports; Doklady; Dynamics of Atmospheres and Ocean; Papers in Applied Geography; Forests; Genesis and Geography of Soils, Eurasian Soil Science; Geography, Environment, Sustainability; Global and Planetary Change; Hydrological Process; International Journal of Biometeorology; International Journal of Climatology;*

Izvestiya, Atmospheric and Oceanic Physics; Journal of American Water Resources Association; Journal of the Atmospheric Sciences; Journal of Climate; Journal of Freshwater Ecology; Journal of Geophysical Research Atmospheres; Journal of Missouri Medicine; Journal of Soil Pedology; Monthly Weather Review; National Weather Digest; The Open Atmospheric Science Journal; Papers in Applied Geography; Pure and Applied Geophysics; Quarterly Journal of the Royal Meteorological Society; Renewable Energy; Tellus A: Dynamic Meteorology and Oceanography; Theoretical and Applied Climatology; and *Transactions of the Missouri Academy of Science.*

The author of 277 conference preprint papers and a contributor to numerous scholarly volumes, he is also the Chief Editor of *Recent Developments in Tropical Cyclone Dynamics, Prediction, and Detection.*

Dr. Lupo is the recipient of the Missouri Academy of Science Most Distinguished Scientist Award, Kemper Teaching Award, School of Natural Resources Excellence in Teaching Award, College of Agriculture Food and Natural Resources (CAFNR) Outstanding Undergraduate Advisor Award, and MU Professor of the Year. He has also been a member of the U.S.-Russian Presidents Climate Subgroup of the Science and Technology Working Group and is a Fellow of the Royal Meteorological Society. In addition, he is a member of the American Meteorology Society; Sigma Xi Honor Society; Phi Kappa Phi Honor Society; Missouri Academy of the Sciences; National Council of Industrial Meteorologists; Fulbright Association; American Geophysical Union; European Geophysical Union; Gamma Delta, Sigma, Society of Catholic Scientists, Honor Society of Agriculture; and National Weather Association.

Independent Institute Studies in Political Economy

THE ACADEMY IN CRISIS | *edited by John W. Sommer*

AGAINST LEVIATHAN | *by Robert Higgs*

AMERICAN HEALTH CARE |
edited by Roger D. Feldman

AMERICAN SURVEILLANCE | *by Anthony Gregory*

ANARCHY AND THE LAW |
edited by Edward P. Stringham

ANTITRUST AND MONOPOLY | *by D. T. Armentano*

AQUANOMICS |
edited by B. Delworth Gardner & Randy T Simmons

ARMS, POLITICS, AND THE ECONOMY |
edited by Robert Higgs

A BETTER CHOICE | *by John C. Goodman*

BEYOND POLITICS | *by Randy T Simmons*

BOOM AND BUST BANKING |
edited by David Beckworth

CALIFORNIA DREAMING | *by Lawrence J. McQuillan*

CAN TEACHERS OWN THEIR OWN SCHOOLS? |
by Richard K. Vedder

THE CHALLENGE OF LIBERTY |
edited by Robert Higgs & Carl P. Close

THE CHE GUEVARA MYTH AND THE FUTURE
OF LIBERTY | *by Alvaro Vargas Llosa*

CHINA'S GREAT MIGRATION | *by Bradley M. Gardner*

CHOICE | *by Robert P. Murphy*

THE CIVILIAN AND THE MILITARY |
by Arthur A. Ekirch, Jr.

CRISIS AND LEVIATHAN, 25TH ANNIVERSARY
EDITION | *by Robert Higgs*

CROSSROADS FOR LIBERTY |
by William J. Watkins, Jr.

CUTTING GREEN TAPE |
edited by Richard L. Stroup & Roger E. Meiners

THE DECLINE OF AMERICAN LIBERALISM |
by Arthur A. Ekirch, Jr.

DELUSIONS OF POWER | *by Robert Higgs*

DEPRESSION, WAR, AND COLD WAR |
by Robert Higgs

THE DIVERSITY MYTH |
by David O. Sacks & Peter A. Thiel

DRUG WAR CRIMES | *by Jeffrey A. Miron*

ELECTRIC CHOICES | *edited by Andrew N. Kleit*

ELEVEN PRESIDENTS | *by Ivan Eland*

THE EMPIRE HAS NO CLOTHES | *by Ivan Eland*

THE ENTERPRISE OF LAW | *by Bruce L. Benson*

ENTREPRENEURIAL ECONOMICS |
edited by Alexander Tabarrok

FAILURE | *by Vicki E. Alger*

FINANCING FAILURE | *by Vern McKinley*

THE FOUNDERS' SECOND AMENDMENT |
by Stephen P. Halbrook

FUTURE | *edited by Robert M. Whaples, Christopher J.
Coyne, & Michael C. Munger*

GLOBAL CROSSINGS | *by Alvaro Vargas Llosa*

GOOD MONEY | *by George Selgin*

GUN CONTROL IN NAZI-OCCUPIED FRANCE |
by Stephen P. Halbrook

GUN CONTROL IN THE THIRD REICH |
by Stephen P. Halbrook

HAZARDOUS TO OUR HEALTH? |
edited by Robert Higgs

HOT TALK, COLD SCIENCE | *by S. Fred Singer*

HOUSING AMERICA |
edited by Randall G. Holcombe & Benjamin Powell

IN ALL FAIRNESS |
*edited by Robert M. Whaples, Michael C. Munger &
Christopher J. Coyne*

JUDGE AND JURY |
by Eric Helland & Alexender Tabarrok

LESSONS FROM THE POOR |
edited by Alvaro Vargas Llosa

LIBERTY FOR LATIN AMERICA | *by Alvaro Vargas Llosa*

LIBERTY FOR WOMEN | *edited by Wendy McElroy*

LIBERTY FOR PERIL | *by Randall G. Holcombe*

LIVING ECONOMICS | *by Peter J. Boettke*

MAKING POOR NATIONS RICH |
edited by Benjamin Powell

MARKET FAILURE OR SUCCESS |
edited by Tyler Cowen & Eric Crampton

THE MIDAS PARADOX | *by Scott Sumner*

Independent Institute Studies in Political Economy

100 SWAN WAY, OAKLAND, CA 94621-1428

For further information:
510-632-1366 • orders@independent.org • http://www.independent.org/publications/books/